丁洁民／主编

TJAD 2012—2017 作品选

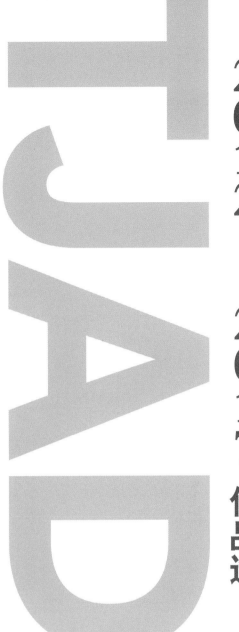

TJAD 2012—2017 作品选

丁洁民 / 主编

广西师范大学出版社
· 桂林 ·

images Publishing

CONTENTS

/

目 录

8   序 / 丁洁民
    Foreword / Ding Jiemin

10 同济设计及其近期作品辨析 / 吴长福
    Analysis of TJAD and Its Recent Works / Wu Changfu

16 上海中心大厦
    Shanghai Tower

CULTURAL AND PERFORMING ARCHITECTURE
26—119
文化建筑

26 上海自然博物馆
    Shanghai Natural History Museum

32 北川地震纪念馆
    Beichuan Earthquake Memorial Museum

38 范曾艺术馆
    Fan Zeng Art Museum

44 山东省美术馆
    Shandong Art Gallery

50 刘海粟美术馆迁建工程
    Liu Haisu Art Gallery

56 2015 米兰世博会·中国企业联合馆
    China Corporate United Pavilion, Expo 2015 Milano Italy

62 扬州广陵公共文化中心
    Yangzhou Guangling Public Cultural Center

68 中国商业与贸易博物馆及义乌市美术馆
    China Business and Trade Museum, Yiwu Art Museum

74 咸阳市市民文化中心
    Xianyang Citizen Culture Center

80 遵义市娄山关红军战斗遗址陈列馆
    Zunyi Loushan Pass Red Army Battle Site Museum

84 长沙国际会展中心
    Changsha International Convention and Exhibition Center

90 上海棋院
    Shanghai Chess Institute

94 吴中区东吴文化中心
    Wuzhong District Dongwu Culture Center

98 福州市城市发展展示馆
    Fuzhou Urban Development Exhibition Hall

102 上海交响乐团迁建工程
    Relocation of the Shanghai Symphony Orchestra

108 嘉定新城保利大剧院
    Jiading New Town Poly Grand Theater

114 武汉电影乐园
    Wanda Wuhan Movie Park

OFFICE, COMMERCIAL AND HOPSCA ARCHITECTURE
120—179

办公、商业、综合体

120	上海市虹口区海南路 10 号地块项目 Plot No. 10, Hainan Road, Hongkou District, Shanghai		150	黄河口生态旅游区游船码头 Cruise Terminal in the Ecological Tourism Zone of the Yellow River Estuary
126	上海市城市建设投资开发总公司企业自用办公楼 Office Building of Shanghai City Construction and Investment Corporation		156	利福上海闸北项目 L-FERG Zhabei Project, Shanghai
132	平凉街道 22 街坊项目 Project of No. 22 Pingliang Street		160	英特宜家武汉购物中心 Wuhan Inter IKEA Shopping Center
136	同济大厦 A 楼 Building A of Tongji Tower		164	温州机场交通枢纽综合体 Wenzhou Airport Traffic Hub Complex
142	中国电子科技集团第三十二研究所 科研生产基地（嘉定） Production Base of the 32nd Research Institute of the China Electronics Technology Group (Jiading)		168	郑州二七新塔 Zhengzhou 27 New Tower
146	英特宜家无锡购物中心 Wuxi Inter IKEA Shopping Center		174	昆明滇池国际会展中心 4 号地块（主塔） Plot No. 4 of Kunming Dianchi International Convention and Exhibition Center (main tower)

SPORTS AND TRANSPORTATION ARCHITECTURE
180—223

体育、交通建筑

180	遂宁市体育中心 Suining Sports Center		208	哈大客专大连北站站房 Station Building of Dalian North Station of the Harbin-Dalian Passenger Line
186	济宁奥体中心 Jining Olympic Sports Center		212	兰州西站站房工程 Station Building of Lanzhou West Railway Station
192	常熟市体育中心体育馆 Gymnasium of Changshu Sports Center		216	海口汽车客运总站 Haikou Coach Terminal
198	上海崇明体育训练基地 Shanghai Chongming Sports Training Base		220	上海吴淞口国际邮轮码头后续工程 Follow-up Project of the Shanghai Wusongkou International Cruise Terminal
204	改建铁路宁波站改造工程 Reconstruction of Ningbo Railway Station			

EDUCATIONAL AND HOSPITAL ARCHITECTURE
224—285

教育、医疗建筑

224 北京建筑大学新校区图书馆
Library in the New Campus of Beijing University of Civil Engineering and Architecture

230 同济大学浙江学院图书馆
Library of Zhejiang College of Tongji University

236 西北工业大学长安校区图书馆
Library of the Chang'an Campus of Northwestern Polytechnical University

240 南开大学津南校区
Jinnan Campus of Nankai University

250 浦江镇江柳路幼儿园
Pujiang Town Jiangliu Road Kindergarten

256 上海市委党校二期工程（教学楼、学员楼）
The School of the Shanghai Municipal Communist Party of China, Phase II (Teaching Building, Student Building)

262 新江湾城中福会幼儿园
New Jiangwan Town Zhongfuhui Kindergarten

268 苏州大学附属第一医院平江分院
The First Affiliated Hospital of Suzhou University, Pingjiang Branch

274 苏州市第九人民医院
The Ninth People's Hospital of Suzhou

278 上海市第一人民医院改扩建工程
Expansion and Reconstruction of the First People's Hospital of Shanghai

282 徐汇区南部医疗中心
Xuhui District Southern Medical Center

HOTEL AND RECONSTRUCTION OF EXISTING BUILDING
286—339

酒店、既有建筑改造

286 青岛岭海温泉大酒店
Qingdao Linghai Springs Hotel

292 国家机关事务管理局第二招待所翻扩建工程
Expansion of the Second Guest House of the National Affairs Management Bureau

298 集美新城万豪酒店
Jimei Marriott Hotel

302 昆明花之城豪生国际大酒店
Howard Johnson Plaza Hotel, Flower City, Kunming

308 黄山元一柏庄国际旅游体验中心国际酒店
International Hotel of Huangshanyuanyi Bozhuang International Tourism Experience Center

312 上海鞋钉厂改建（原作设计工作室）
Reconstruction of the Shanghai Hobnail Factory (Original Design Studio)

318 同济大学设计创意学院
College of Design and Innovation, Tongji University

324 延安中路816号（解放日报社）
No. 816 Yan'an Middle Road (Jiefang Daily)

330 同济大学博物馆
Museum of Tongji University

336 同济大学建筑与城规学院D楼改建项目
Reconstruction of Building D of the College of Architecture and Urban Planning, Tongji University

DATA OF SELECTED WORKS
340—349

附录：作品数据

FOREWORD

/

序言

丁洁民 /
同济大学建筑设计研究院（集团）有限公司总裁

依托百年学府同济大学的深厚底蕴，同济大学建筑设计研究院（集团）有限公司（以下简称 TJAD）是一所集学术研究、工程实践和人才培养等优势于一体的大型综合类设计研究院。经过近 60 年的锤炼与积累，TJAD 从品牌知名度、社会责任、企业规模、技术实力及经济效益等各方面领跑，成为了国内建筑设计公司的排头兵。

研究与设计并重是高校设计院区别于一般设计院的特点之一，也是 TJAD 一直坚持的宗旨。我们坚信并贯彻着建筑创作的"先导"价值，整个企业上下形成了重视建筑创作的整体氛围。将建筑创作过程看作一种工作的习惯和必然，让建筑师成为文化的载体，在每一次创作中都能够从文化本质中汲取能量，提供最合适的解决策略，同时强大的技术研发和集成能力又成为实现建筑创意的坚实保障。正是基于这样的坚持，我们才能在行业的快速变化发展中，积极主动地进行创作，真正的去思考如何避免浮躁、切实提升工程质量。

当面临着国内建筑设计行业市场转型、竞争格局改变等重要战略机遇时，市场的各个层面对于设计的要求越来越"苛刻"，这也推动着我们以更为理性和严谨的设计态度来应对市场的变化。从本册作品集所收录的项目中可以看出，过往的五年内，我们的项目类型更加多元化、项目技术更加集成化、项目设计更加精细化。TJAD 传统强项文化建筑充分体现了设计创作中的文化引领作用——我们既积极参与了国际国内的"大事件"，交出了像米兰世博会中国企业联合馆、北川地震纪念馆这样的焦点建筑；我们同时也设计出大量真实且关联城市文化发展的建筑，积极地为地方建设贡献智慧，使得建筑呈现出纷繁的多样性，以求真正推动所在区域的文化交流，探索可持续发展的可能性，如山东美术馆、范曾艺术馆、刘海粟美术馆、上海棋院、吴中区东吴文化中心等项目就是其中的代表作品。

本书也同时收录了 TJAD 一直关注的办公、商业、体育、交通、教育、医疗、酒店和既有建筑改造等方面的项目，以便更为全面的解读与回顾集团近五年的发展。它们都在一定程度上展现了过去五年中 TJAD 在全国各地的大型建筑中所发挥出的巨大的辐射能量以及笃定有声的建筑态度。

此外，我们也保持着与各大国际知名建筑事务所的紧密合作关系，随着上海中心、上海自然博物馆、上海交响乐团迁建工程及保利大剧院等项目的建成使用，充分证明了 TJAD 有能力整合不同领域的优秀资源和先进技术，为所有的参建者提供良好的合作平台，为每一座城市的地标性建筑建造保驾护航。

我们以此作品的集结出版作为对上一个五年的过往成就的小结，一方面是希望能较为完整的呈现 TJAD 的创作水准、技术能力及价值导向；另一方面也是给予我们自己以前进的动力和保持脚踏实地精心设计的工作作风，并呈现出更多能够唤起文化沉思的建筑作品。保持行业的引领地位是 TJAD 不懈努力的方向，希望我们的设计作品能够为每一个城市带去新的活力，并且通过这些作品，让 TJAD 的设计精神传播到更为广阔的交流平台，助力行业进一步发展。

Established on the framework set by Tongji University's century of accumulated cultural knowledge and expertise, Tongji Architectural Design (Group) Co., Ltd. (TJAD) is a large-scale integrated design institute bringing together academic research, engineering, and talent cultivation. Through nearly sixty years of growth and effort, TJAD has become a leader in social responsibility, business scale, technical capabilities, and economic results; thus becoming a leader among domestic architectural design companies.

Placing equal importance on research and design is one of the features that distinguishes TJAD from ordinary design colleges, and is the tenet at the core of TJAD's works. We firmly believe in implementing pioneering ideas in our architectural creations and strive to create an atmosphere thoroughly focused on architectural creation. From senior management to junior staff, all employees must regard the architectural creation process as habitual and a working necessity, and to regard architects as indispensable vessels of culture. We absorb energy and inspiration from the surrounding culture of every creation to make the most appropriate strategic solutions possible. Meanwhile, our powerful technical R&D and coordination skills provide a solid base that allows our architectural creativity to flourish. It is because of our insistence on our perpetual evolution within the industry, and our resolution to engage in a proactive and positive creative process, that we are able to avoid mediocrity and improve our engineering quality in a practical way.

Confronted with the strategic opportunity for market transformation and changes in the competitive framework of the domestic architectural design industry, we have learned to cope with market dynamics with a positive, reasonable, and rigorous attitude that takes the increasingly "harsh" requirements imposed on the design process from different aspects into account. This collection of works suggests that in the past five years, TJAD has undertaken an increasingly diversified range of projects using more integrated technologies and more refined designs. Among these, the cultural buildings, which are the traditional strength of TJAD, fully realize the inspirational power and potential for cultural evolution inherent in design and creation. We have actively participated in major architectural events both domestic and international, and constructed a number of influential buildings, such as the China Corporate United Pavilion for the Expo Milano, and the Beichuan Earthquake Memorial Museum. We have also designed a large number of critical and interconnected urban cultural structures, making our contribution to local development and diversifying architectural aesthetics to promote regional culture exchange, and to explore possibilities for sustainable development. For instance, the Shandong Art Museum, Fan Zeng Art Gallery, Liu Haisu Art Gallery, Shanghai Chess Institute, and Wuzhong District Dongwu Culture Center are all representative works of this philosophy in action.

In addition to these projects, TJAD has also engaged in the construction of offices, businesses, sports facilities, transportation hubs, education facilities, medical complexes, hotels and historical building reconstruction projects, which have been collected in this book to comprehensively interpret and review the last five years of TJAD's development. These representative works display the enormous range and influence that TJAD's major architectural projects nationwide over the last five years.

Beyond these physical works, TJAD has maintained close cooperative relationships with renowned architectural firms around the world. The construction and operation of the Shanghai Tower, the Shanghai Natural History Museum, the Shanghai Symphony Orchestra relocation project, and the Poly Grand Theater are powerful demonstrations of TJAD's ability to integrate the resources and advanced technologies of various sources into a range of projects, offering a good platform for cooperation for all partners, and engaging in the construction of landmark buildings for every city TJAD engages in.

This book is a summary of our company's past five years, and aims to give a comprehensive presentation of the Tongji Institute's creations, techniques, and values, as well as to inspire us to continue engaging in practical and elegant design projects. We wish to present an increasing range and number of culturally inspirational structures to the world. Our mission is to maintain a leading position in the industry and our wish is to inject new vitality to every city to host our designs. Through these projects, we have been able to promote TJAD's design spirit on a broader platform and boost the industry's development.

ANALYSIS OF TJAD AND ITS RECENT WORKS

同济设计及其近期作品辨析

吴长福 / Wu Changfu

同济设计近期代表作品又如期亮相。一个设计机构的阶段性业绩，既依赖于整个社会经济的发展水平与发展形势，也反映出一个阶段内设计机构自身的运行状态与对产品质量的追求。同济设计作品是依托同济设计集团长期以来的管理、技术与文化积累，针对特定的社会需求，通过设计人员艰苦的专业性劳动所完成的创作成果。它们既烙有同济设计的历史印痕与特质，又透射着超越以往的新探索与时代感，从而呈现出一种在传承中不断进取的整体品牌风貌。

设计作品的阅读，即便专心致志、穷尽精力，仍是对于事物表象的关注。我们虽然可从中搜寻到许多精彩的设计语汇，但无法了解其深层的、多重意志博弈下的设计生成过程。然而，一旦把众多设计作品看作是一个互相联系的整体，便可捕捉到一些线索，并足以借此去认识一个设计团体于个体之上的集体设计取向与作为。从同济设计近期代表作品的群像中，可由自我完善的纵向维度与行业比较的横向维度，来体会同济设计对于隐涵于作品之中的若干重要辩证关系的出色把握。或许，这正是一个优秀设计机构的身份所现、实力所为与成功所在。

其一，多元化与专业化。

在设计创作上，特色的凝练是设计机构集聚自身优势、提升市场竞争力的必要举措。同济设计集团作为一家综合性大型设计企业，它的特色就是全面性、多元化，即无论应对设计领域中的何种需求，都能提供全方位优质的设计服务与解决方案，这也正是一个大型企业的必然选择。这种无所偏颇、能从容应对设计市场的基础，除了来自长期的技术积淀外，主要是依托于同济大学雄厚的学术背景。学校在设计相关领域学科齐全、阵容强劲，为同济设计提供了源源不断的专业思想、专业知识与专业人才的支持，是同济设计得以全面发展的重要技术后盾。

多元化首先体现在设计项目类型构成的多样性。此次选辑的作品，包含5个部分、10种设计类型的60个项目。从大规模的办公商业综合体、交通、体育建筑，到中、小型的文化、教育建筑，作品几乎涵盖了所有建筑功能类型，以及新建、改建、保护等建设类型。其次是设计思路或创作途径与表现手法的丰富性。就作品所呈现出的多元特征而言，很难能用类似于何种风格的词语来一言蔽之。当然，在一个个特定的作品里，我们依然能感受到不同的思想火花的闪烁，以及共同释放出的智慧光泽。（图1）

多元化是一种集体策略，而专业化则是确保具体项目高质量完成的关键。综合设计机构不等同于拥有全能的建筑师或工程师。通过有效的设计管理与组织，发挥设计人员与设计团队的技术专长，进行针对性的设计投入，是对社会与业主的一种高度负责态度，也是优秀作品诞生的基本前提。在同济设计近期作品中，经验效应与技术含量显而易见，一些项目的主创设计人员长期从事相关领域的设计与研究工作，他们本身就是国内该设计领域中一流的专家学者。多元化的发展构架与专业化的技术配备，共同确立了同济设计的专业地位与社会声誉。（图2）

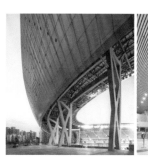

1   2

吴长福 / 同济大学建筑设计研究院（集团）有限公司副总裁，同济大学建筑与城市规划学院教授

Wu Changfu / Vice President of Tongji Architectural Design (Group) Co., Ltd., and Professor of the College of Architecture and Urban Planning of Tongji University

1 规模逾 45 万平方米的郑州二七新塔超高层项目与面积不到 4500 平方米的同济大学博物馆改造项目
Zhengzhou super high-rise 27 New Tower (over 450,000m²) and Tongji University Museum (less than 4,500m²) reconstruction project

2 专业技术支撑下的大跨度建筑项目：遂宁市体育中心与铁路宁波站改造工程
Large-span architecture with professional technical support: Suining Sports Center and the Ningbo Railway Station Reconstruction Project

Tongji Architectural Design (TJAD) has recently exhibited another batch of recent works. The progress and regular improvement of a design institute not only relies on the surrounding society's level of development, but also reflects the design institute's operating philosophy and its pursuit of quality. TJAD's works are the creative achievements of its designers and their specialized labor, and were developed for specific social needs under the long-term management and technical and cultural talent development of TJAD. These works have been imprinted with the historical legacy and characteristics of TJAD, yet carry a sense of wonder and exploration and timeliness, presenting a progressive brand image that reflects the institute's heritage.

Experiencing design through literature is always fundamentally focused on the superficial image of the object, no matter how concentrated and devoted the reader is to the work's deeper meaning. Although we may encounter a rich and expressive design vocabulary in the described works, we fail to understand the subtleties of design process with its concomitant conflicts and contradictions. However, as long as we regard multiple works as part of an interconnected whole, we may glimpse some clues to help us understand the overarching philosophy of the design and the choices of the design team. The images recently released by TJAD allow us to understand several important dialectical relationships concealed in their works, from reaching new heights and more audacious designs of in its vertical structures. Perhaps, this is exactly how an ideal design institute should act, and reflects TJAD's strength and shows the reasons for its success.

First, diversification and specialization.

Offering a dense and varied package is vital for design institutes to create market advantages and improve their competitive standing. As a major comprehensive design company, TJAD offers both a comprehensive and a diverse range of services. In other words, it is able to provide an outstanding well-rounded design service that offers solutions no matter what request is put forward. This is a necessity for any major design company. However, the ability to approach all angles of the design market with consistency comes from the long-term accumulation of technical expertise, as well as from the support of the rich academic background of Tongji University. Class after class of design majors and esteemed faculty of Tongji University have offered their professional opinions, professional knowledge, and professional talent to TJAD. Their expertise has been a vital technical support for TJAD's all-round development.

TJAD's diversification is first and foremost presented in its diversified range of design projects. The works selected in this collection include sixty projects across ten design categories in five parts. They range from large office buildings and commercial complexes, to transportation and sports facilities, to medium and small scale cultural and educational buildings. This collection covers almost all functional types of architecture, including newly-built, renovations, and preservation projects. Secondly, there is a great richness and variety in both design concepts and in creative approaches and presentation methods. It is very difficult to summarize the diversified characteristics of these works in any brief or movement-centric stylistic vocabulary. However, in each specific work we can feel the sparks of inspiration and the glory of wisdom released from the piece. (Figure 1)

Diversification is an overarching strategy, while specialization is the key to ensure the high quality of individual projects. Creating an integrated design institute does not necessarily mean simply hiring a genius architect or engineer, because it multiplies the technical strengths of designers and their design teams through effective design management and organization. It arranges the inputs purposefully with respect to the design, fulfilling a strong responsibility to the society and the owner, as well as previously established conditions for the creation of outstanding works. TJAD's recent works have show clear growth in experience and technical skills. The main designers of some of these projects have long been engaged in the design and research of their fields of focus. They are first-class scholars and experts in architectural design. The combination of a diversified development structure and refined and modern techniques have set the bar for TJAD and built its national reputation. (Figure 2)

3 具有传统空间意蕴与地域特征的范曾艺术馆与浦江镇江柳路幼儿园
　Fan Zeng Art Gallery and Pujiang Town Jiangliu Road Kindergarten: traditional space implications and regional features
4 与境外著名建筑师的合作项目：嘉定新城保利大剧院与上海市虹口区海南路10号地块项目
　Cooperation with famous foreign architects: Jiading New Town Poly Grand Theater and Shanghai Hongkou District's Hainan Road Plot No.10
5 基于环境关系与建筑性格表达的山东省美术馆与南开大学津南校区大学生活动中心
　Shandong Art Gallery, the Student Activity Center of Nankai University Jinnan Campus: an expression of environmental relationships and architectural personality

其二，地域性与开放性。

建筑存在于具体的场所之中，其地域属性毋庸置疑。在社会经济发展趋于一体化的背景下，地域性问题集中体现在如何对待地理环境与传统文化方面。由此，关注建筑的地域性，也就反映着建筑师对于具体建造环境与文化的尊重。同济设计自20世纪50年代起，从未间断对建筑地域性表达的探索实践，在当下的设计中更有着充分地自觉流露。这些作品中所强调的建筑地域性，不仅在形式意象上有所体现，更多地贯穿于空间构成、材料运用、节点处理以及色彩选择等诸多方面，展现出基于特定建造地域与功能要求的全新建筑形象。（图3）

地域具有的边界特征决定了地域性存在积极一面的同时，也带有局限性。以更为开放的姿态来积极应对包括地域性表达等在内的一系列建筑创作问题，是同济设计不断前行的内生动力。海纳百川、大气谦和——上海的城市精神也滋养着同济整个建筑学科博采众长的专业氛围。掌握国内外行业最新发展动态，汲取专业最新发展成果，开展最为广泛的学术理论与设计实践交流，使同济设计创作始终拥有宽阔的国际视野。近年来，特别是在与国际顶尖设计机构和设计师的合作方面，又有了切实的推进，从之前的技术配合、到同场竞技、再到各展所长的合作设计，同济设计以扎实精湛的整体设计水准，在与国际建筑设计同行的同步发展中，彰显出了自身的非凡价值。（图4）

其三，创新性与完成度。

建筑设计能谓之建筑创作，其核心是创新，这种创新是基于建筑设计目标的创新性作为，它可以贯穿于设计过程与设计成果的各个环节。追求设计创新作为一种价值导向与群体意志，它决定了同济设计的创作特色与创作文化。在同济设计近期作品中所表现出的种种原创性探索，均是围绕具体项目各自的关键性问题来展开的，因此具有真实性，且绝无形式上的造作感。例如，通过对基地特征与城市肌理的深入分析而形成的与环境有机协调的独特建筑姿态；通过对交通流线的定性、定量梳理而提供的多重空间组织方案；通过对纷杂功能要素的归并整合而创造出的简洁又富有感染力的空间形象，等等。对材料、建造以及生态等技术的研究深化，更为同济设计创新开辟了无限的空间。（图5）

3

4 5

Second, regionalism and openness.

Since buildings are situated in a specific place, they inherently acquire the geographical attributes of that place. Against the background of simultaneous social and economic development, the problem of regionality deals with the question of how to treat the geographical environment and the local historical culture. Therefore, a concern with regional architecture is a responsibility of the architect to the site's environment and its local culture. TJAD has never stopped its exploration and practice of regional expression of architecture since the 1950s, and this quality has been thoroughly and consciously reflected in current designs. The regional architectural styles emphasized by these works, in addition to shapes and images, is applied more to spatial structure, material application, node disposal, and color selection, etc., reflecting a new architectural image based on specific regions and the individual structure's functional requirements. (Figure 3)

The boundary features of the region determine both the advantages and limitations of the site. TJAD's core strength is its ability to forge ahead in its projects by coping with the architectural creative process with a more open mind, which embraces regional expression. The inclusive, generous, and humble spirit of Shanghai has inculcated in TJAD the idea of learning from strengths of others. They regularly master the latest dynamics of the industry in China, absorb the best ideas of the most advanced professionals and their latest developmental achievements, and carry out the most extensive academic theoretical works and debates on practical designs, and these activities all broaden TJAD's international horizons for their design creations. In recent years, especially in cooperation with top international design institutes and designers, TJAD has made astonishing progress—it has achieved synchronous development with international peers in technical cooperation, competition and design cooperation, highlighting the institute's extraordinary value. (Figure 4)

Third, innovation and completion.

Why do we call the product of architectural design an architectural creation? It is because architectural design is led by the motivational core of innovation; it is an innovative accomplishment based on the goal of architectural design. It connects all links between design process and design achievement. The pursuit of innovative design, as a kind of value orientation and the collective will, determines the creative features and culture of TJAD. The experimentalism and originality revealed in TJAD's recent works has been centrally oriented around the key problems of each project, and are reflected in real and tangible design patterns. For instance, a given building's unique architectural shape harmoniously coordinated with the environment can only be achieved after an in-depth analysis of the site's features and the urban surroundings. A multi-spatial organization solution can only be developed after qualitative and quantitative analysis of traffic flow; a concise yet transcendental spatial image is only created after a summary and integration of complicated functional elements. The ever-expanding research on materials, construction methods, and technologies has opened an infinite realm for TJAD to continue its innovation process. (Figure 5)

建筑设计的创新性并不意味着对经验的抛弃,它是对经验的重新审视与修正,是经验的升华。同济设计长期的技术储备与经验积累,保证了在强调创新的同时,依然使其设计作品具有高品质的完成度。可以认为,同济设计的创新性亦包含着对建筑最终完成度的刻意要求。在我国当下正处在设计供需关系多变、设计质量亟待全面提升的背景下,同济设计以其宏观的创意追求与精到的细节把控,用作品垂范,为整个设计行业增添了积极的正能量。(图6)

其四,经济效益与社会效益。

设计作为一种产业,它处于城市物质空间建设产业链的上游,其经济性作用不可小觑。设计机构作为从事设计的主体,对于经济利益的追逐,是企业生存与可持续发展的根本需要。同济设计在与大多数设计机构一样看重设计产品数量与规模的同时,更注重产品质量所能产生的附加效应。设计项目可有大小,设计题材可有易难,设计内容可有主次,但对设计创作要求绝无差别,质量标准比量齐观。这种看似缺乏经济性选择的设计投入,其实赢得的正是业主与市场的信任,以及随之而来的经济效益。(图7)

设计是智力劳动,更是社会性服务。在经济效益之上的是设计的社会效益,也就是设计对社会进步与发展的贡献。同济设计秉承百年同济"聚焦国家发展战略、服务经济社会发展"的郑重承诺,主动对接社会诉求。在近期作品中,除了大量性的城市开发项目外,还可以看到针对教育、医疗建筑等关乎国计民生的公共服务设施的设计探索,针对生态节能、既有建筑改造等国家建设重点问题的设计方法与技术应用实践,针对在国家与地区政治、经济生活中的重大事件,以及应对突发灾害等建设难点问题所开展的设计参与工作。同济设计以专业智慧与热情肩负起了一个高校设计机构的社会责任与担当。(图8)

6 细部设计与建造完成精良的上海棋院与上海自然博物馆
Shanghai Chess Institute and the Shanghai Natural History Museum with their thoroughness and exacting detail in design

7 突出功能效益的长沙国际会展中心与注重创意表现的 2015 米兰世博会中国企业联合馆
Changsha International Convention and Exhibition Center and China Corporate United Pavilion, Expo 2015 Milano, highlighting its functional advantages and focusing on creative expression

8 具有重要社会意义的娄山关红军战斗遗址陈列馆与北川地震纪念馆
Loushan Pass Red Army Battle Site Museum and Beichuan Earthquake Memorial Museum, with their important social significance

Embracing innovation in architectural design does not mean to ignore experience. Innovation is the review and modification of experience, as well as the sublimation of experience. The long-term technical reserves and experiential accumulation within TJAD have ensured that the institute can construct works with high quality in addition to maintaining a degree of originality and innovation in each piece. We may admit that the innovation of TJAD has been forced to set its limits with the final completion of each building. In China, given the vast differential between design supply and demand, and the urgent demand to improve design quality, TJAD has injected positive energy to the entire design industry with its works, setting a strong example with its macroscopic creative pursuit and precise control of details. (Figure 6)

Fourth, economic benefits and social benefits.

Design is placed at the forefront of the urban spatial construction industrial chain, and the economic benefits of these projects should not be overlooked. The design institute, as a central agent of the design process, chases after economic benefits, because it is a fundamental requirement for the survival of the business and of sustainable development. While most other design institutes attach great importance to the quantity and scale of its design products, TJAD has paid more attention to the additional concern of quality. There are big and small design projects with either easy or difficult design themes, with varied or more monotonous sets of design contents, but the requirements set on design and creation are universally the same. The quality standards are the same. This design input, though seeming to lack economic choices, is actually the key to win trust from the owners and the market, and the economic benefits that come along with it. (Figure 7)

Design is both an intellectual effort and a social service. Before its economic benefits comes the social benefit of design. In other words, design is fundamentally about the contribution of design to social progress and development. TJAD, upholding the solemn promise of "focusing on the national development strategy and serving social and economic development" of the hundred-year-old Tongji University, takes the lead to respond to social needs. In addition to a number of urban development projects, we see creative exploration of education and medical buildings and other critical public facilities, practical applications and experimental design methods in eco-architecture, energy efficiency, building restoration, and other key national construction issues, as well as designs for national and regional political, economic, and social events, as well as architectural and engineering responses to recent natural and humanitarian disasters in their recent works. TJAD shoulders all the social responsibilities and duties of a major design institute with professional wisdom and passion. (Figure 8)

SHANGHAI TOWER
上海中心大厦

Located in the finance and trade zone of Lujiazui, Pudong District, the main building of Shanghai Tower has a height of 580 meters and a total height of 632 meters, and covers an area of 30,000 square meters. The Shanghai Tower focuses on business, offering venues for conventions and exhibitions, hotels, sightseeing and entertainment, commerce, and so on. The Tower has five functional areas, the public entertainment area, the lower, central and upper office areas, the corporate clubhouse area, the boutique hotel area, and the elite functional experience area.

总平面图 / Site plan

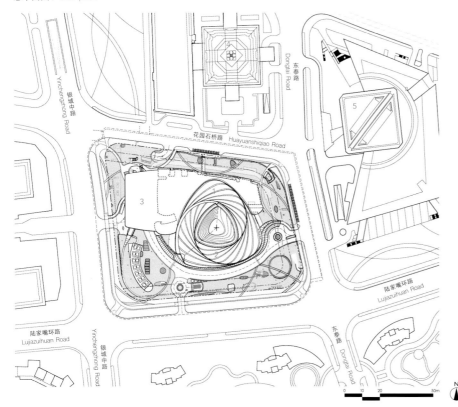

1 上海中心大厦 / Shanghai Tower
2 塔楼 / Tower 126F
3 裙楼 / Podium 7F
4 金茂大厦 / Jinmao Tower
5 上海环球金融中心 / Shanghai World Financial Center

01 眺望上海中心 / Distant view
02 从豫园处远眺上海中心大厦 / Distant view from Yuyuan Garden
03 下沉庭院 / Sunken courtyard

04　上海中心大厦 / Shanghai Tower

上海中心大厦位于上海浦东的陆家嘴功能区，占地3万多平方米，主体建筑结构高度为580米，总高度632米。项目以办公为主，其他业态有会展、酒店、观光娱乐、商业等。大厦分为五大功能区域，包括大众商业娱乐区域，低、中、高办公区域，企业会馆区域，精品酒店区域和顶部功能体验空间。

本项目采用了双层表皮的设计，是世界最高的双层表皮建筑。"上海中心"内部由九个圆形建筑彼此叠加构成，其间形成九个垂直空间。大厦双层表皮的内层覆盖了垂直的内部建筑，而三角形外幕墙则形成第二层表皮，而外立面与内里面之间的空间则形成了空中中庭。"上海中心"把这一系列的垂直社区用于不同的用途，每个垂直社区均设计为独立的生物气候区进行调节和使用，这不仅可以改善大厦内的空气质量，还能降低"烟囱效应"，创造宜人的休息环境。而这些公共活动楼层给里面上班的人员提供了休闲的场所，可以减少大厦用户上下楼梯的次数。

大厦幕墙也是可持续性的绿色设计，与传统的直线形建筑相比，大厦内部圆形立面使其眩光度降低了14%，而且减少了对能源的消耗。大厦螺旋形顶端可以收集雨水，进行回收利用。螺旋形不对称的顶端处理还可以降低大厦的风载，顶部安装的风力涡轮发电机还能为大厦提供绿色电能。

大厦打造绿色建筑的理念，采用多项最新的可持续发展技术，达到绿色环保的要求。是国内第一座按照中国绿色建筑评价体系和美国LEED绿色建筑认证体系设计的"绿色摩天大楼"。

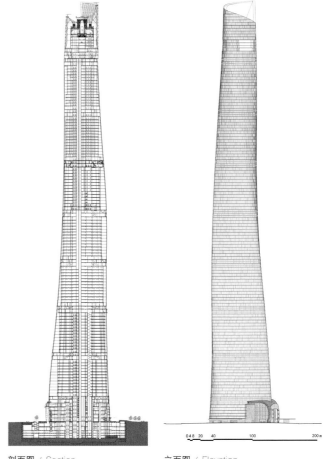

剖面图 / Section　　立面图 / Elevation

05 屋顶花园 / Roof garden
06 双层幕墙 / Double curtain wall

一层平面图 / First floor plan

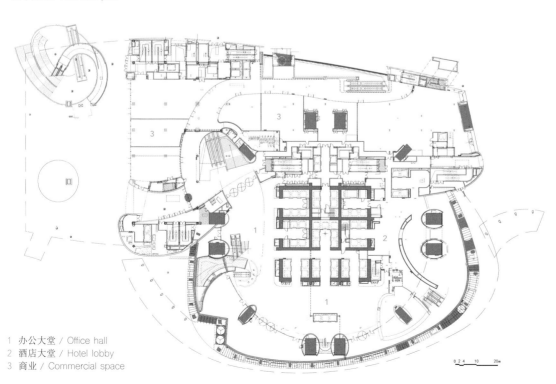

1 办公大堂 / Office hall
2 酒店大堂 / Hotel lobby
3 商业 / Commercial space

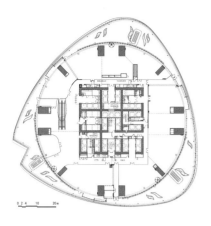
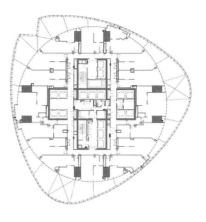
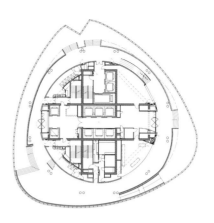

办公大堂平面图 / Office hall plan　　酒店平面图 / Hotel plan　　观光厅平面图 / Sightseeing hall plan

07　上海中心大堂 / Hall of Shanghai Tower
08　观复宝库半亩园 / Roof garden 'Banmuyuan'
09　观光层 / Sightseeing floor
10　多功能厅 / Multi-function hall

11 地下通道 / Underground pass way

SHANGHAI NATURAL HISTORY MUSEUM
上海自然博物馆

Located in Jing'an District, the Shanghai Natural History Museum is at the junction of Shanhaiguan Road and Datian Road, and neighbors Jing'an Sculpture Park. Built on 12,000 square meters of land with a gross floor area of 45,000 square meters, the museum has three floors above ground and two underground. Below the basement lies a station for the city's Metro Line 13, which was built synchronously with the museum above. The tower has a maximum height of 18 meters and the basement has a depth of 22 meters. The museum mainly serves the function of exhibition, supplemented by education, social interaction, and natural experience.

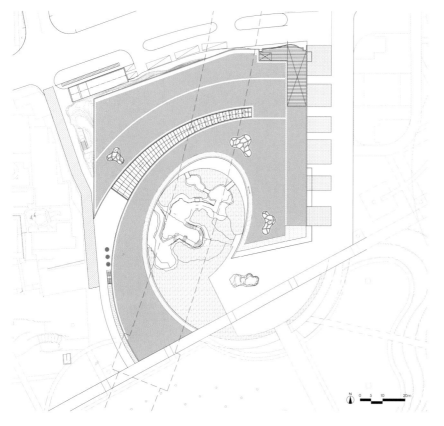

总平面图 / Site plan

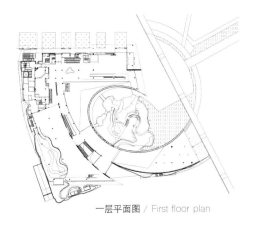
一层平面图 / First floor plan

二层平面图 / Second floor plan

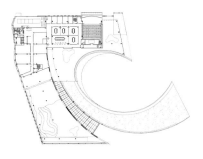
三层平面图 / Third floor plan

01 总体鸟瞰效果图 / Rendering of overall aerial view

上海自然博物馆位于上海市静安区，山海关路、大田路交界处，临静安雕塑公园。建筑用地面积 1.2 万平方米，总建筑面积 4.5 万平方米，地上三层、地下两层。地下室下方为地铁十三号线区间路段，两者整体建构，地上总高度 18 米，总埋深 22 米。博物馆主要功能为展览，兼具教育、社交和自然体验的功能。

建筑的整体形态设计源于绿螺的壳体形式，螺旋上升的绿色屋面从雕塑公园内升起，地面提升至 18 米的高度，并围合出一面椭圆形中央景观庭院，使之成为贯穿整个建筑的参观流线所围绕的中心焦点。

博物馆的大部分面积是展厅，其他功能包括管理人员办公、周转库房、停车，以及为参观者服务的餐厅、商店，还有一个可独立开放的 Imax 影院。这些功能 75% 被安排在地面以下，使地面上的建筑体量大幅减小以融入雕塑公园的绿化中。

建筑南立面为巨大的弧形细胞状玻璃幕墙，总高度 33 米，无结构柱，由人类细胞结构的图案构成。细胞墙内外层次丰富，跨越地下及地上各层，本身已成为一道令人震撼的景观，并为博物馆地下空间带来了良好的采光。

博物馆在建筑设计、展陈设计各方面均代表了行业发展的最高水平，采用了多项最新的可持续发展技术，获得了中国绿色建筑三星级认证和美国 LEED 金奖，将绿色建筑与建筑空间表现完美地融合，起到极佳的示范效应；建筑与地铁轨道线的结合设计，解决了一系列震动、噪音的难题；高技术的展陈手段和大量古生物真迹展示相结合，使上海自然博物馆在业内达到领先水平。

西侧立面 / West elevation

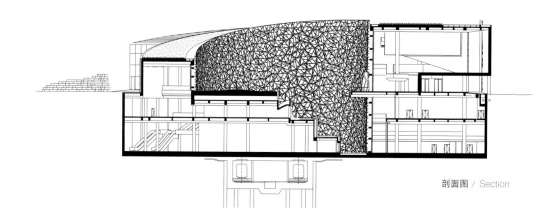

剖面图 / Section

02 南向透视 / South perspective
03 下沉庭院鸟瞰图 / Aerial view of the sunken courtyard

04　展陈区透视图 / Perspective of the display area
05　入口门厅坡道透视图 / Rampway perspective of the entrance hallway
06　室内中庭透视图 / Perspective of the interior courtyard

BEICHUAN EARTHQUAKE MEMORIAL MUSEUM

北川地震纪念馆

The Beichuan Earthquake Memorial Museum is a memorial site built to remember the loss of the May 12 Wenchuan earthquake, with building functions including memorial of the central event, artifact display, science popularization, education, and scientific research. The museum takes up an area of 14.23 hectares, and contains a main hall and a side hall with a combined floor area of over 14,000 square meters. This is the only national memorial museum built in honor of the Wenchuan earthquake in 2008, and is a part of Beichuan Earthquake Ruins Park.

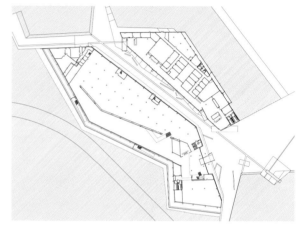

一层平面图 / First floor plan

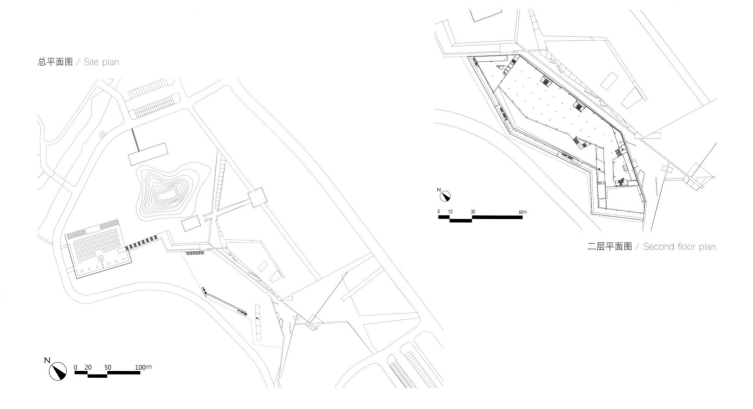

总平面图 / Site plan

二层平面图 / Second floor plan

01　纪念馆入口广场鸟瞰图 / Aerial view of the memorial entrance plaza

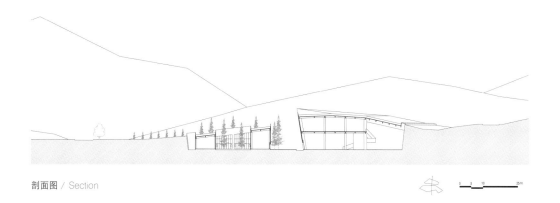

剖面图 / Section

北川地震纪念馆是在"5.12"汶川大地震历史背景下以留存灾难记忆、展现人类大爱以及人对自然的再认识为规划目标，确定了纪念馆的总体定位——一个人类历经特大地震灾难的纪念性遗址纪念馆，主要包括纪念、展示、科普、教育、科研五大功能。

纪念馆选址北川曲山镇任家坪，涵盖北川中学遗址，东临山东大道。整个馆区占地约14.23公顷。纪念馆共分为主馆与副馆两部分，建筑面积超过1.4万平方米。该项目是2008年汶川大地震后修建的唯一的国家级纪念馆，属于整个北川地震遗址公园的组成部分之一。

纪念馆的设计从"裂缝"这一地震瞬间留在大地的永恒印记获得启示，以大地艺术的造型策略再现"裂缝"，并以此为空间主导要素串联起整体园区与建筑的重要节点，形成连续、完整的参观、悼唁路线。设计中，作为园区重要部分的北川中学遗址通过大地艺术的语言得以保留，使后人能永远记住这场灾难以及逝去的生命。象征性与美学语言、简单抽象与通俗可读、内敛与纪念性三组关系成为建筑与景观视觉控制的关键，以此传递尊重自然、与自然和谐相处的设计理念。

建筑师通过放弃地面突出建筑形象的选择表达一种对自然之力的思考与尊重和对蔑视自然行为的反思；通过"裂缝"中充满力量的暗红色锈蚀钢板与自然环境形成鲜明对比，将人们带回山崩地裂的瞬间，营造出逼真、艺术、震撼的纪念空间效果；通过"裂缝"整合纪念园参观流线，为参观者提供了一条从不同距离、不同高度、不同尺度感受遗址的纪念之路，以强烈的空间体验带来巨大的精神震撼。

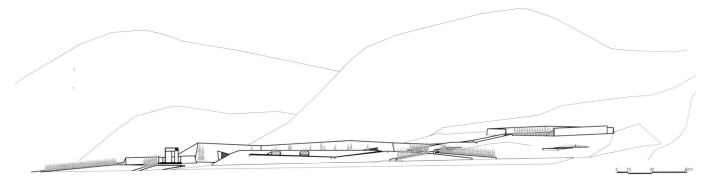

山东大道立面图 / Elevation of Shandong Avenue

02 纪念馆主馆主立面 / Elevation of the main memorial
03 纪念步道 / Memorial walk
04 祭奠园望向"裂缝" / 'Crack' view from memorial garden

05　纪念馆主馆门厅 / Hall of the main memorial
06　纪念馆主馆楼梯 / Stair of the main memorial
07　祭奠园 / Memorial garden

FAN ZENG ART GALLERY
范曾艺术馆

The Fan Zeng Art Gallery was specially built to display, exchange, study, and collect the paintings, calligraphy, and artworks of master Fan Zeng, along with the poems and prose of the Fan family. Starting with the traditional "courtyard" space, the Fan Zeng Art Gallery harmoniously integrates the visiting process with an emotional experience to create a sense of "understanding the ancients and reflecting the modern heart."

总平面图 / Site plan

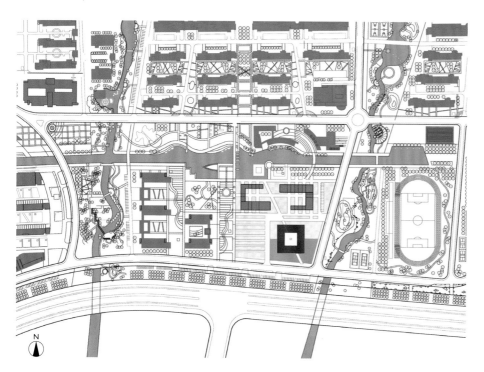

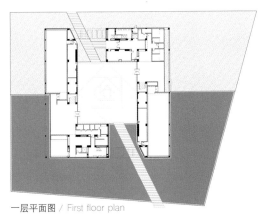

一层平面图 / First floor plan

二层平面图 / Second floor plan

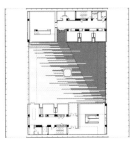

三层平面图 / Third floor plan

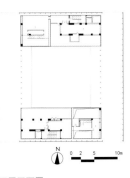

四层平面图 / Fourth floor plan

01　南立面全景 / View of south elevation

范曾艺术馆是为满足范曾大师书画艺术作品以及南通范氏诗文世家的展示、交流、研究、珍藏的需要而建造的。

范曾艺术馆以传统的空间"院"为切入点，将院落从物化关系中脱离，继而呈现游目与观想的合一，以达到"得古意而写今心"的意境。

范曾艺术馆强调的"关系的院"，首先表现在同时呈现的三种不同的院落形式：建筑底层的"井院"、建筑二层南北穿通的"水院"与"石院"、建筑三层四边围合的"合院"。并在此基础上构架起以井院、水院、石院、合院为主体的叠合的立体院落。"叠合院落"的初衷是期望在受限的场地上化解建筑的尺度，将一个完整的大体量化解为三个局部的小体量，这便于我们以身体尺度完成对院落的诠释。我们从类似于网格化的控制体系以及整合全局的大秩序中脱身出来，从局部的略带松散的关系开始。就像看似不相干的三种院子，由于各自的生长理由被聚在一起，也因连接方式的不同而出乎意料地充满变数。

范曾艺术馆强调的"观想的院"，以局部关系并置的方式形成时间上的先后呈现，为"游目"式的观想体验提供可能。在非同时同地的景物片段中，局部的关系先后呈现。

范曾艺术馆强调的"意境的院"，讲求"计白当黑"的意境、有无之间的把控、与不饱满中呈现饱满的观想。它没有设定一个强大的整体框架，将所有的情节归入明晰的主线索之中。而是依照三种不同院落的自发性生成秩序铺展开略带松散的局部关系。

范曾艺术馆如同一个可以水墨浑融的空灵腔体，为浓进淡出的晕染留有发挥的余地。它是向水墨致敬的一种态度，方寸之物，内有乾坤，于局部的单纯中体悟整体的复杂，单纯澄净而又气韵饱满。

02 鸟瞰图 / Aerial view

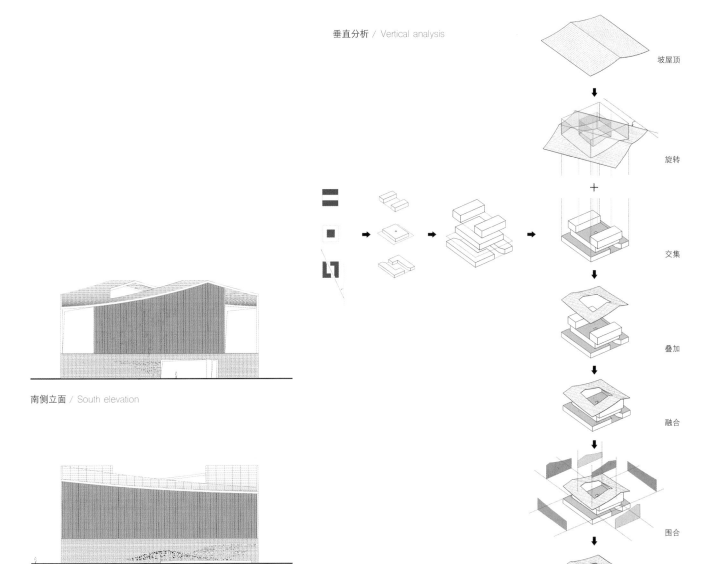

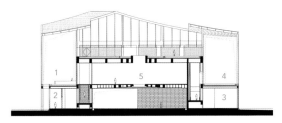

剖面图 -1 / Section-1

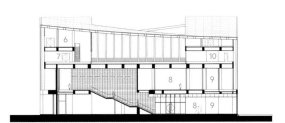

剖面图 -2 / Section-2

1　石院 / Stone courtyard
2　入口门厅 / Entrance hallway
3　展厅 / Exhibition room
4　水院 / Water courtyard
5　主展厅 / Main exhibition room
6　会议室 / Meeting room
7　研究室 / Research room
8　休息区 / Resting area
9　电梯厅 / Elevator hall
10　陈列区 / Display area

03 藻井局部 / Part of the caisso
04 二层水院 / Water courtyard on the second floor
05 三层屋顶合院 / Roof courtyard on the third floor

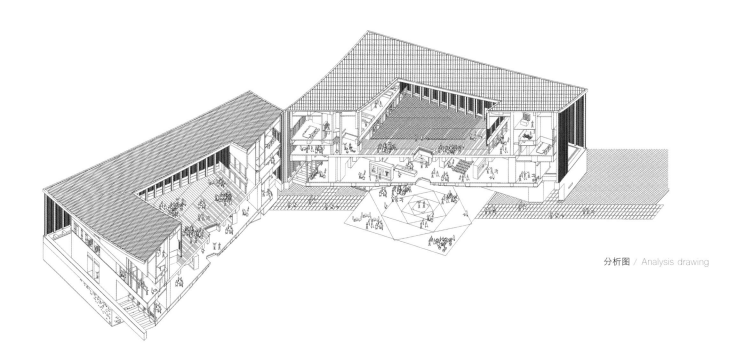

分析图 / Analysis drawing

SHANDONG ART GALLERY
山东省美术馆

The Shandong Art Museum is an ultra-large exhibition building and the largest modern art museum in China. Constructed on the central idea of integrating multiple urban spaces, the design of Shandong Art Museum positively engages and draws inspiration from the surrounding environment.

总平面图 / Site plan

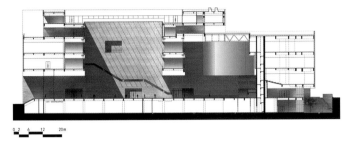
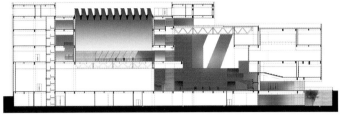

剖面图 / Section

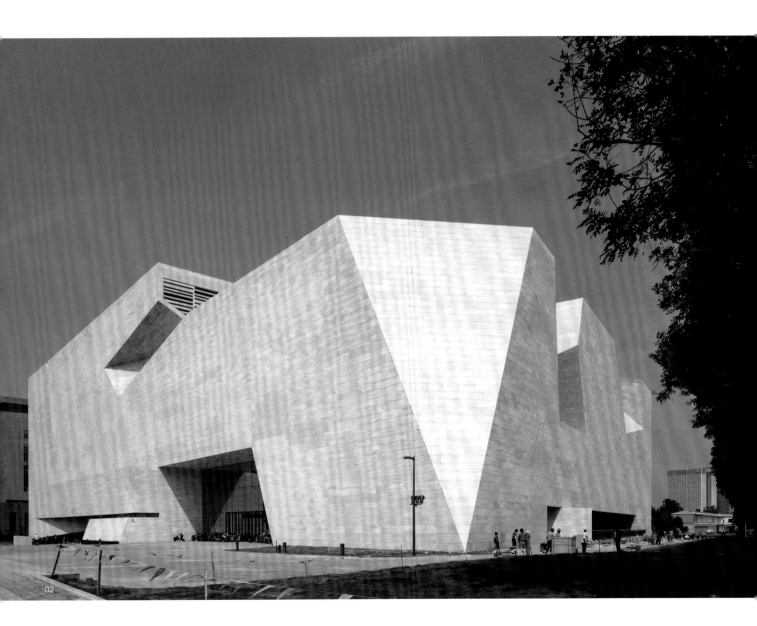

01 主入口 / Main entrance
02 建筑外景 / Exterior view

山东省美术馆属于特大型博览建筑，是我国新建的规模最大的现代美术馆。山东省美术馆总体设计立足于城市空间整合的理念，对周边城市环境做出积极响应。建筑设计植根于特定的场地条件和山东深厚的历史人文内涵，建筑布局合理回应功能要求，依据大型美术馆的复杂功能要求完善功能配置，尤其是代表大型美术馆特点的货运设施、流线安排、备展空间以及照明设施的设计周详。建筑形体呈现为正在渐变中的形态——具有"山形"特征的建筑形体逐渐过渡到方整规则的状态。室内设计是建筑概念的延续，以"山"为主题的中央大厅和以"城"为主题的二层大厅相互交融、对比统一。内部空间设计以视线分析为基础，空间界面层叠错落展开。自然采光与空间布局紧密结合，打造出理想的当代艺术展示空间。

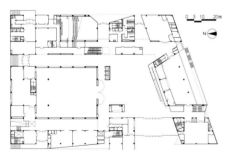
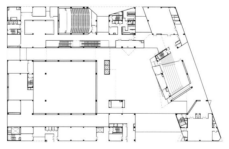

一层平面图 / First floor plan　　　一层半平面图 / 1.5 floor plan

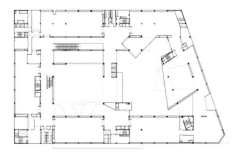
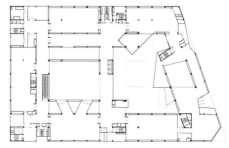

二层平面图 / Second floor plan　　　三层平面图 / Third floor plan

03 南立面外观 / South elevation
04 东立面局部 / Part of east elevation
05 中央大厅全景 / Central hall

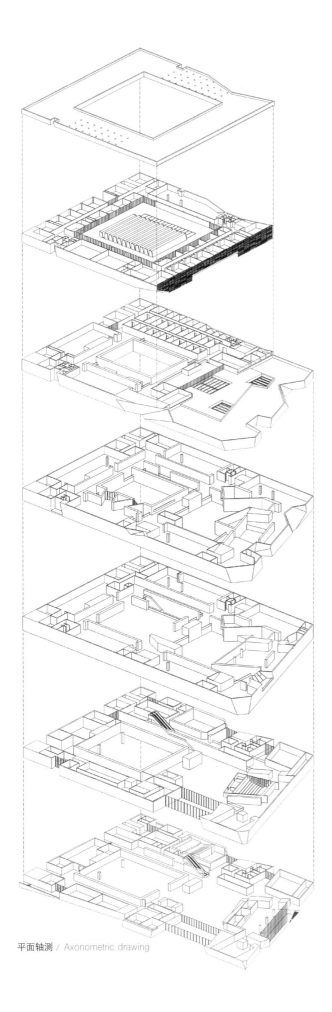

平面轴测 / Axonometric drawing

06 阶梯教室 / Lecture hall
07 二层公共大厅 / Public hall on the second floor
08 中央大厅仰视 / Bottom view of central hall

06

LIU HAISU ART GALLERY
刘海粟美术馆迁建工程

The relocation of the Liu Haisu Art Gallery was a major cultural construction project in Shanghai, with the new site taking up an area of 6,000 square meters. The above-ground building has a total height of 23 meters and a gross floor area of 12,540 square meters, with three floors for the art gallery, museum and Liu Haisu memorial hall. It's a comprehensive museum that meets the design standards of central national art museums.

1 主出入口 / Main entrance
2 下沉庭院出入口 / Entrance of the sunken courtyard
3 货运出入口 / Freight entrance
4 非机动车库出入口 / Entrance of non-motor garage
5 地下车库出入口 / Entrance of the underground garage
6 员工出入口 / Staff entrance
7 公共通道 / Public pass way
8 场地主出入口 / Main entrance of the site
9 场地人行出入口 / Pedestrian entrance

总平面图 / Site plan

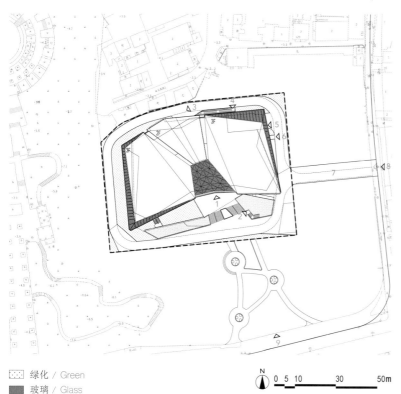

绿化 / Green
玻璃 / Glass

1 下沉庭院 / Sunken courtyard
2 序厅 / Lobby
3 内库房 / Exterior storage
4 机械式停车库 / Mechanical garage
5 常设展厅 / Permanent exhibition room
6 露台 / Terrace
7 设备平台 / Facilities terrace
8 公共资料阅览室 / Public reading room
9 艺术配套服务 / Art supporting service
10 坡道 / Ramp

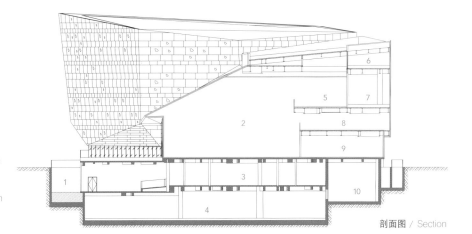

剖面图 / Section

01　鸟瞰图 / Aerial view

刘海粟美术馆迁建工程是上海市重大文化建设项目，总用地面积6000平方米，地上三层地下两层，高度23米，总建筑面积约12540平方米，是集美术馆、博物馆和刘海粟个人纪念馆功能于一体的综合场馆，设计标准为国家重点美术馆。

刘海粟美术馆迁建工程的设计从刘海粟深厚的人生与艺术积淀中汲取灵感，设计取意其不拘一格、激情豪气的人生态度及艺术气质，以耸立的形体、倾斜的中庭和大气的入口呼应原美术馆的造型，通过大手笔的体量切割塑造出强烈的雕塑感。

美术馆的设计立意为"云海山石"，取意于刘海粟一生"为师为友"的黄山。山浮云海之上，高洁轻灵，既是国画永恒的主题，亦是东方传统艺术中抽象审美哲学的精神内核。飞扬的建筑实体如浮于云海之上的黄山山石，峻峭而飘逸。富有力度感的折面勾勒出大气的建筑体量，亦为建筑赋予强烈的动感和冲击力。中庭玻璃天窗的分隔与走向沿袭中国古典建筑的椽与屋檐，将东方艺术气息与充满诗意的静谧融入美术馆。现代与传统、东方艺术与西方美学有机地融合，既是刘海粟一生的艺术成就，也是刘海粟美术馆的设计理念。

美术馆以中庭为核心，主流线与公共空间有机融合，将交通区域与不同的展厅联成起伏变化的流动空间。建筑平面功能布局上分区明确，动静明晰，将美术馆的功能予以最大的延伸。

刘海粟美术馆迁建工程地面层及以上为钢结构，并充分发挥钢结构的特点——着重体现设计的雕塑感体量，以及丰富那些极富变化的展示空间。设计始终在功能、形式与专业的设备技术要求中寻求平衡，在如悬挑结构、富有挑战性的幕墙设计，以及极为紧凑的平面布局等设计处理中充分考虑相关因素，使之达到效率的最优化。

02 西南侧日景 / View from southwest
03 南侧立面夜景 / Night view of the south elevation

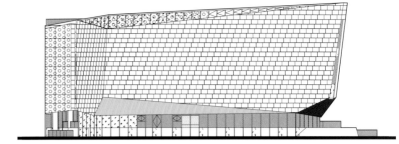

西侧立面 / West elevation

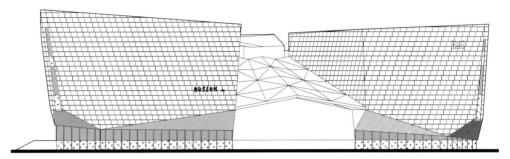

南侧立面 / South elevation

一层平面图 / First floor plan

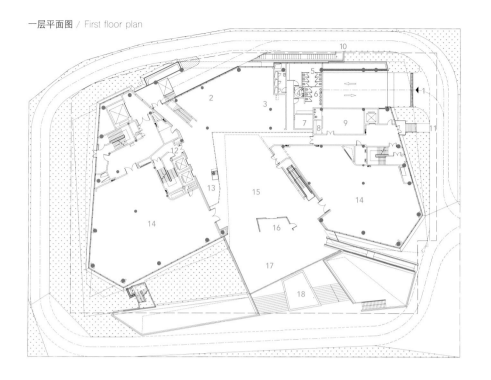

二层平面图 / Second floor plan

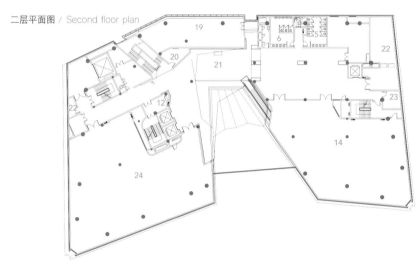

三层平面图 / Third floor plan

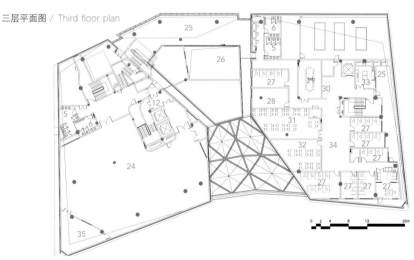

1 地下非机动车库入口 / Entrance to underground garage for non-motors
2 茶座 / Teahouse
3 艺术配套服务 / Art supporting service
4 无障碍厕所 / Barrier-free toilet
5 女卫生间 / Lady's room
6 男卫生间 / Men's room
7 行李寄存室 / Left baggage
8 茶水间 / Tea room
9 贵宾室 / VIP room
10 地下机动车库出入口 / Entrance to underground garage for motor vehicles
11 员工、VIP入口 / Entrance for staff and VIP guests
12 前室 / Prechamber
13 领票问询导览 / Ticket inquiry and guide
14 临时陈列厅 / Temporary showroom
15 序厅 / Lobby
16 安检区 / Security check
17 美术馆主入口 / Main Entrance to gallery
18 花池 / Flower bed
19 设备平台 / Facilities terrace
20 母婴室 / Maternal infantile war
21 公共走道 / Public passway
22 准备间 / Preparation room
23 服务间 / Service room
24 常设陈列厅 / Permanent showroom
25 露台 / Terrace
26 常设陈列厅上空 / Overhead of permanent showroom
27 行政办公 / Office of administration
28 数据机房 / Data room
29 会议室 / Meeting room
30 前台 / Reception
31 研究档案室 / Archives of research
32 艺术设计室 / Art and design room
33 接待室 / Reception office
34 休息区 / Rest area
35 模拟画室 / Simulated studio

04 入口透视 / Perspective of the entrance
05 二层中庭 / Courtyard on the second floor
06 中庭大厅 / Hall of the courtyard

CHINA CORPORATE UNITED PAVILION, EXPO 2015 MILANO

2015 米兰世博会·中国企业联合馆

Located in the NE.6 sector of Expo 2015 Milano, the China Corporate United Pavilion is to the north of Expo Avenue and east of the landscape axis. Taking up an area of 1,270 square meters, it has three floors above ground with a total height of 12 meters and nearly 2,000 square meters of floor area. The use of two-dimensional contrasts like "square and round," "internal and external," and "hard and soft" were adopted to demonstrate traditional Chinese philosophy to visitors.

总平面图 / Site plan

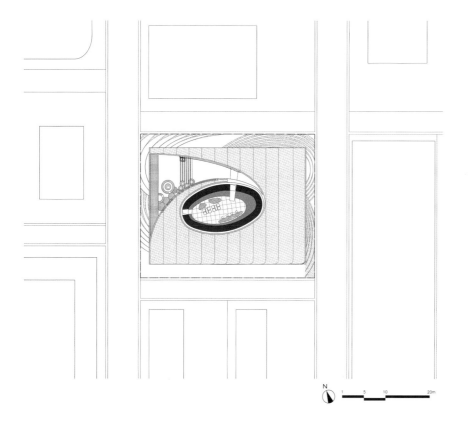

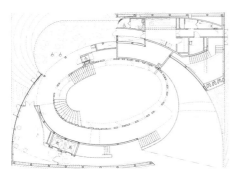
一层平面图 / First floor plan

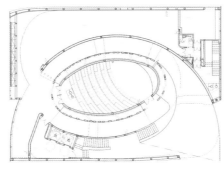
二层平面图 / Second floor plan

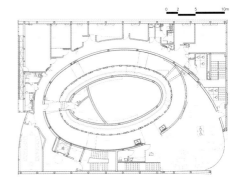
三层平面图 / Third floor plan

01 半鸟瞰日景 / Aerial view

中国企业联合馆位于2015年米兰世博会NE.6地块，东西向世博大道以北、南北向景观轴以东。项目占地1270平方米，地上三层，建筑面积约2000平方米，建筑高度12米。

世博会的主题"滋养地球，生命的能源"呼吁人类正视"索取与反哺"的平衡。中国传统思想中"反者道之动"的认识体现了古人面对矛盾关系时善于平衡的智慧。我们采用"方圆""内外""刚柔"等一系列二元构成的建筑手法，使观众领会中国传统思想的意境。

基于场地的正交网格形态，建筑的基本体量为方形，我们在其中插入椭圆形"绿核"、柱筒、中庭及环形坡道。在广泛存在于大自然中的螺旋形式里巧妙地组织内外空间的转换，在室内空间中纳入绿化、木质、空气、阳光等自然元素。

建筑贴临基地边界，为了塑造符合公共建筑的入口尺度，在12米总高的三层小建筑中切出一个净高7.5米的空间，几重坡道均展露于此，并将其与外部场地空间融合，产生"小中见大"的效果。

幕墙采用钢板材料，通过钢板的延展性完成对复杂曲面的塑造。钢板可焊接、打磨、再喷漆，能消除立面的分缝，与选择膜材料同样成为简化立面元素的手段。膜材料的触感和视觉感受，也是柔中带刚的二元关系的体现。

建筑主要竖向受力构件为内圈的树状柱筒和外层的立面桁架。树状柱筒是结构设计中的关键部位，也是建筑形态的重要元素。高效的立面桁架同时遵从建筑轮廓，使建筑外表皮向上掀起，划出一道与地面相切、直至屋面的大弧线。

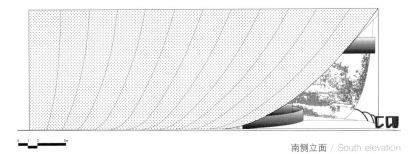
南侧立面 / South elevation

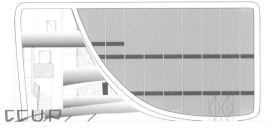
东侧立面 / East elevation

02 南侧立面透视图 / Perspective of south elevation
03 东侧立面透视图 / Perspective of east elevation

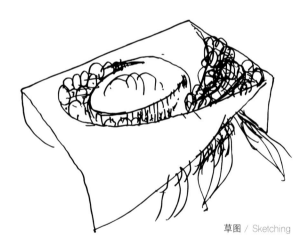

草图 / Sketching

剖面图 / Section

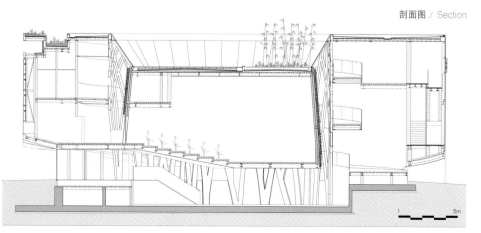

04 坡道内部透视图 / Interior perspective of the ramp
05 主入口内部透视图 / Interior perspective of the main entrance
06 屋顶花园透视图 / Perspective of the roof garden

YANGZHOU GUANGLING PUBLIC CULTURAL CENTER
扬州广陵公共文化中心

Located in Guangling New Town, Yangzhou City, the center's plot has an area of 45,320 square meters hosting a large, comprehensive public cultural complex, including the Guangling District Library, a culture hall, and public service areas, as well as exhibition and show rooms, commercial buildings, dining, parking lots, and other ancillary services. The gross floor area reaches 97,500 square meters.

总平面图 / Site plan

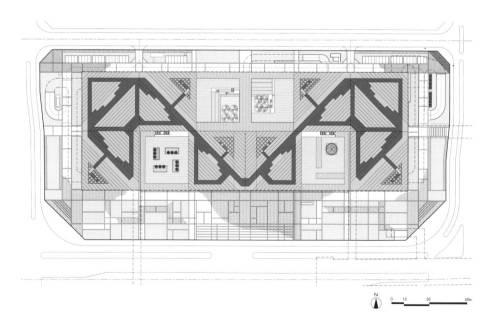

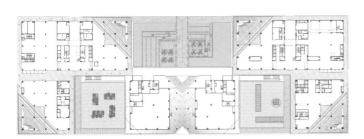

一层平面图 / First floor plan

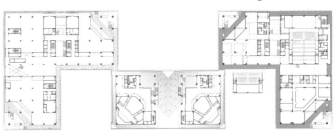

二层平面图 / Second floor plan

01 鸟瞰图 / Aerial view

02 南立面透视 / Perspective of south elevation
03 东南侧效果图 / Rendering of southeast elevation

项目位于扬州市广陵新城，地块总用地面积为45320平方米。使用性质为综合性大型公共文化建筑综合体，包含广陵区图书馆（以下简称图书馆）、文化馆功能，配套市民服务功能以及展演、商业、餐饮、停车等衍生服务功能，总建筑面积为97500平方米。

广陵公共文化中心采用了舒朗隽永的建筑布局与形象，是对城市场所特征的建筑解读。设计师梳理构成周边城市的相关要素，使总体布局建构保持城市空间秩序的延续性，构建统一的城市空间和视觉体系；通过错位相扣的布局形式整合公共开放空间、城市界面、景观、视线等诸多要素，形成独特的城市空间体验；空灵飘逸的一体化形态设计凸显建筑组群整体特征，提升文化中心的辐射力、影响力和吸引力。

为了应对城市功能诉求的建筑表述，设计师打造了一个多元互动的精神家园：以服务型图书馆为核心，多元文化活动为外延，体验互动型文化衍生产业为周边的公共生活平台和市民精神家园；以日常性介入的态度，有意模糊建筑室内外空间的界限，将城市公共活动自然引入建筑内部；以多样性的路径组织建筑内部的公共交通体系，将地下、空中、屋顶的步行体系串联整合，形成弥漫性的探索体验。

被打造成古意今心的立体园林的广陵公共文化中心，也是对扬州城市特色文化的一种体现："地缘文化"的时代演绎，以当代空间构成方式和建构技术阐释本土园林文化的特征；以"立体、弥散、体验"的现代新园林空间融入建筑群的空间序列，丰富了文化中心的漫游体验；"多元化"的内部园林空间塑造，应对不同使用功能的性格特征，形成丰富而有机的景观路径系统。

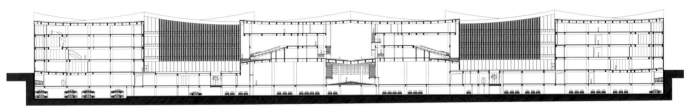

剖面图 / Section

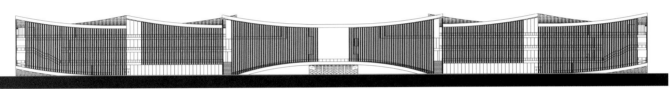

立面图 / Elevation

03

65

04 立体园林 / Interior vertical garden
05 冬潭院 / Dongtan Courtyard
06 下沉广场 / Sunken plaza

05

06

气泡图 / Bubble diagram

CHINA BUSINESS AND TRADE MUSEUM, YIWU ART MUSEUM
中国商业与贸易博物馆及义乌市美术馆

Located in the eastern part of the Yiwu International Culture Central Area, the project sits opposite to the world-renowned Yiwu Trade City with the rising new CBD across the river. Possessing a gross planned area of 66,000 square meters, the museum is located on the western end of the plot with an area of about 42,000 square meters, while the art museum is located on the east of the plot, with an area of about 14,000 square meters.

总平面图 / Site plan

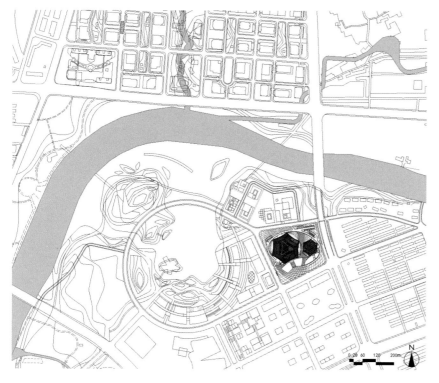

01 沿江透视图 / Perspective along the riverside

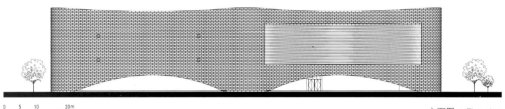

立面图 / Elevation

项目位于义乌市国际文化中心区东部，北临义乌江，与世界知名的义乌商贸城和崛起中的新城CBD隔江相对。规划总用地6.6万平方米，其中博物馆用地位于西侧，占地约4.2万平方米，美术馆用地位于用地东侧，占地约1.4万平方米。

作为国际文化中心区核心节点，商贸博物馆应融合江山呼应的宏观环境形态与尺度，在地脉与地景的层面确立其标志性。以义乌地区的山水为灵感，呼应自然，创造了连绵起伏、舒缓流畅的建筑形态。作为当代中国商贸文明的代表，义乌具有举世闻名、独树一帜的特有商业文化，设计以义乌市花——"月季花"为造型来源，结合义乌"商贸之都"的向心性、辐射性等特点，归纳出"盛世绽放——广宇六合"的布局理念，将义乌市独特的商业文化特征融合其中。

基于当地气候的热力学研究，风的流动成为塑造建筑形式的主要因素，不同层次的"热力学庭院"是整个公共空间系统的核心，从下沉庭园、核心中庭、观景庭园、季节庭园到屋顶景观平台，形成立体而有节奏的空间系统。

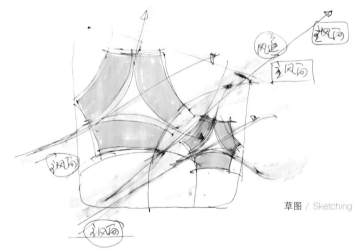

草图 / Sketching

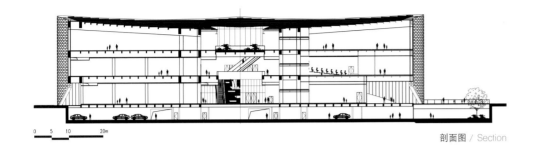

剖面图 / Section

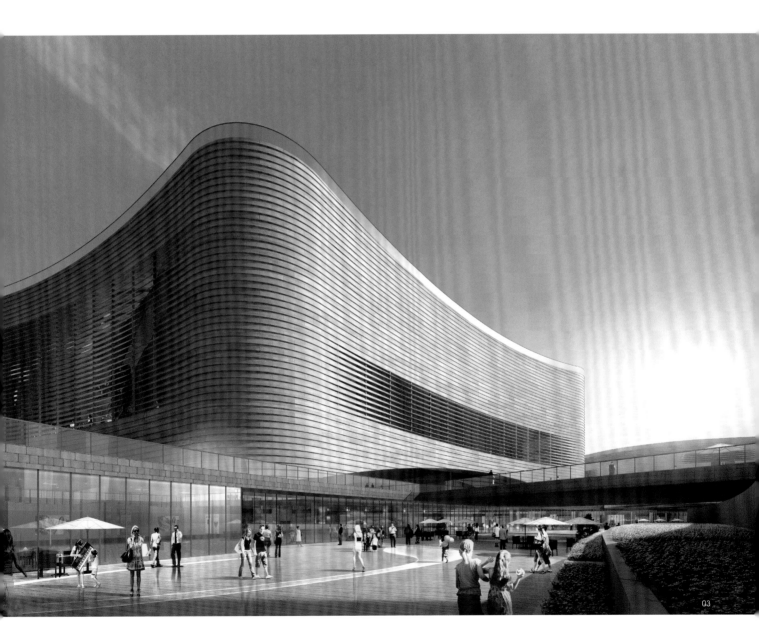

02 总体鸟瞰图 / Overall aerial view
03 下沉广场透视图 / Perspective of the sunken plaza

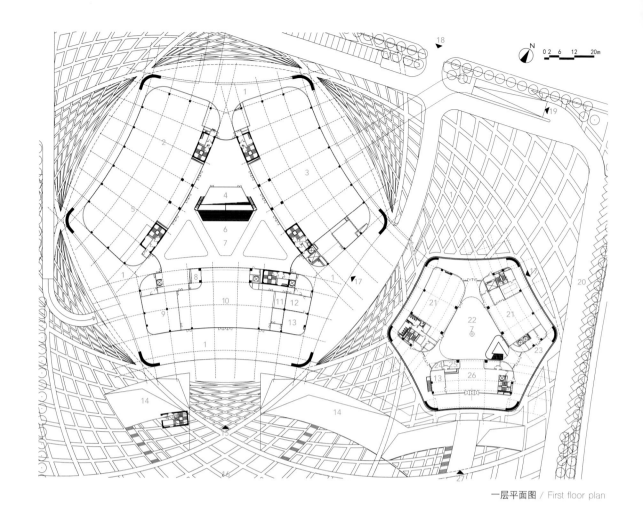

一层平面图 / First floor plan

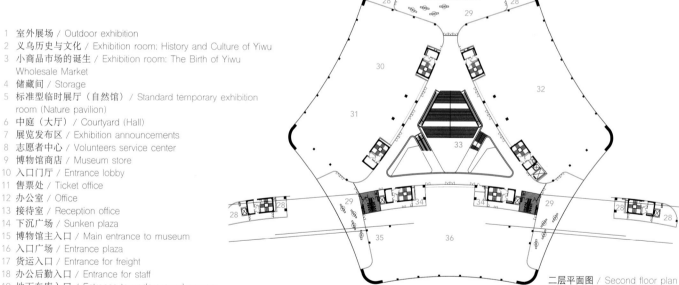

二层平面图 / Second floor plan

1 室外展场 / Outdoor exhibition
2 义乌历史与文化 / Exhibition room: History and Culture of Yiwu
3 小商品市场的诞生 / Exhibition room: The Birth of Yiwu Wholesale Market
4 储藏间 / Storage
5 标准型临时展厅（自然馆）/ Standard temporary exhibition room (Nature pavilion)
6 中庭（大厅）/ Courtyard (Hall)
7 展览发布区 / Exhibition announcements
8 志愿者中心 / Volunteers service center
9 博物馆商店 / Museum store
10 入口门厅 / Entrance lobby
11 售票处 / Ticket office
12 办公室 / Office
13 接待室 / Reception office
14 下沉广场 / Sunken plaza
15 博物馆主入口 / Main entrance to museum
16 入口广场 / Entrance plaza
17 货运入口 / Entrance for freight
18 办公后勤入口 / Entrance for staff
19 地下车库入口 / Entrance to underground garage
20 地面停车场 / Parking lots
21 临时展厅 / Temporary exhibition hall
22 大厅 / Hall
23 画室出入口 / Entrance to studio
24 服务室 / Service room
25 管理室 / Management room
26 门厅 / Lobby
27 美术馆主入口 / Main Entrance to gallery
28 空调机房 / Air conditioning facilities
29 敞开式外廊 / Open veranda
30 可变型临时展厅 / Alterable temporary exhibition room
31 标准型临时展厅 / Standard temporary exhibition room
32 中国古代商业贸易 / Commercial trade in ancient China
33 中庭上空 / Overhead of courtyard
34 讲解员办公室 / Guide office
35 当代贸易体验区 / Experience area for modern trade
36 当代中国商业贸易 / Exhibition room: Commercial Trade in Contemporary China

04 二层通用展厅透视图 / Perspective of the exhibition hall on the second floor
05 室内中庭透视图 / Perspective of interior courtyard

XIANYANG CITIZEN CULTURE CENTER

咸阳市市民文化中心

Situated in the start-up area of Xianyang Beiyuan New Town, to the west of Xianbei Avenue and north of Planning Road, the Xianyang Citizen Culture Center includes four functional areas over 155,000 square meters. The design began with an interpretation of local history and the culture, as well as a detailed analysis of the site.

总平面图 / Site plan

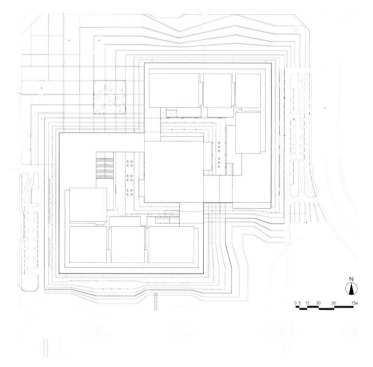

剖面图 / Section

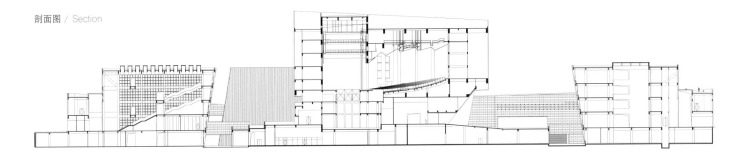

01 鸟瞰图 / Aerial view

02 透视图 / Perspective
03 立面效果图 / Rendering of elevation

咸阳市市民文化中心位于咸阳市北塬新城起步区，咸北大道以西，规划路以北，建筑面积 155000 平方米，包含四大功能片区。

设计从对历史文化的解读和对场地条件的分析两方面着手：

对历史文化的解读：作为中国第一个统一封建制国家的帝都所在，书同文、车同轨……咸阳集中国文化之大成，充分体现了中华文化交融和开放的博大胸襟。

对场地条件的分析：本案位于北塬新城起步区，紧邻五陵塬历史文化景观带。基地东西两侧的东湖和南湖景观区是五陵塬在起步区的景观延伸。因此，本案必须超越自成体系的传统景观模式，着眼于更大范围的城市区域，完成城市区域内的完型。

在以上解读的基础上，本项目总体构思归纳成以下三方面：

勾连围合——应对场地特征的建筑解读：梳理城市架构要素，追求空间秩序的延续性；整合四大片区功能要素，加强相互之间的有机联系；凸显建筑组群整体特征，提升文化展示的辐射力；塑造建筑组群内部空间，增加文化活动的吸引力。

行合趋同——应对地域文化的建筑表述：汲取"大同"的文化思想，体现雄浑大气的建筑意蕴；传承"开放"的文化态度，营造张弛有度的文化氛围；融入"传统"的建筑语汇，表达立足本土的建筑情怀；结合"创新"的时代精神，创建具有东方韵味的时代巨作。

合而不同——应对文化建筑的特殊需求："地缘文化"的时代演绎，以当代科技阐释本土文化；"地标形象"的重点塑造，激发本土文化的认同感；"互动体验"的空间构成，提升文化活动的亲和力；"本位关系"的深入探讨，提升文化活动的凝聚力。

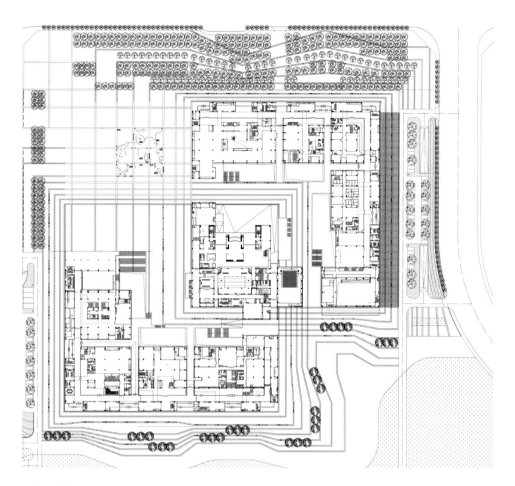

一层平面图 / First floor plan

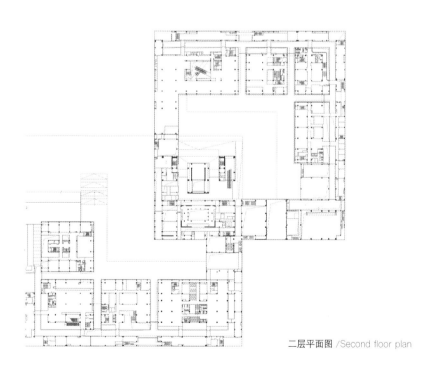

二层平面图 / Second floor plan

04 室外景观透视图 / Perspective of outdoor landscape
05 透视图 / Perspective
06 室内透视图 / Interior perspective
07 透视图 / Perspective

ZUNYI LOUSHAN PASS RED ARMY BATTLE SITE MUSEUM
遵义市娄山关红军战斗遗址陈列馆

Loushan Pass, known as Taiping Pass, is located at the junction of Zunyi and Tongzi, and is an important pass between Sichuan and Guizhou Provinces. This is a new museum to review the Red Army's battle at Loushan Pass. It includes an introduction hall and three theme halls, an auditorium with 200 seats, and administrative office, and its gross floor area reaches 6,056 square meters. It has one floor above ground and one underground, and the building above ground has a height of 14.5 meters.

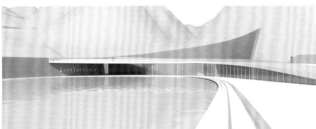

灰模 / Plaster model

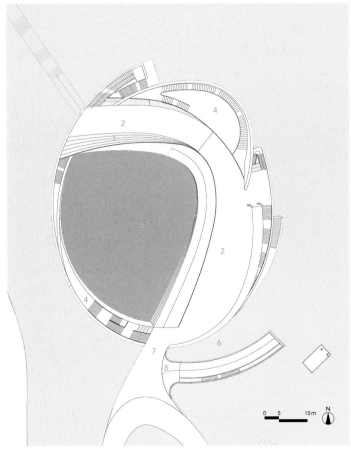

总平面图 / Site plan

1 种植屋面 / Planting roof
2 上人屋面 / Accessible roof
3 台阶 / Steps
4 下沉庭院 / Sunken courtyard
5 景观水池 / Landscape pool
6 景观绿化 / Green landscape
7 广场入口 / Entrance to plaza
8 地下车库入口 / Entrance to underground garage

01 鸟瞰图 / Aerial view

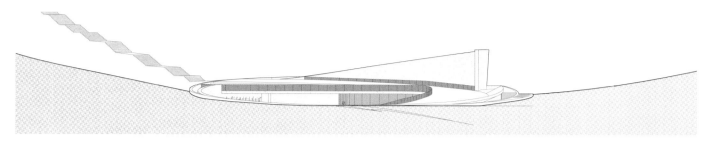

南侧立面 / South elevation

娄山关亦称太平关，位于遵义、桐梓两县交界处，是川黔交通要道上的重要关口。本项目是一座关于红军娄山关战役的新陈列馆，包括一个序厅和三个主题展厅，以及一个200座的报告厅和辅助的办公管理用房，总建筑面积6056平方米。建筑地上一层，地下一层。建筑高度14.5米。

坐落在娄山关山脚的这座红军战斗陈列馆，掩映在崇山峻岭之间。在这里，大自然是一切的统领。因此设计师选择了将建筑物的主要体量都埋在地下，建筑本身以一种大地景观的姿态在此扎根，以地景式建筑"去建筑化"的策略来应对这个特殊环境中的特定体裁建筑的营造。

当年红军万里长征，千回百折，在取得了这场关系着中央红军生死存亡战斗的胜利之后，促使毛主席在此场战役之后写下了《忆秦娥·娄山关》这首大气磅礴的诗词。战役是这所陈列馆所要呈现的主题，在本案中，我们希望以一种对比的方式纪念这场标志性战役的胜利——越是宁静的气氛，越发衬托出当年战争场面的刀光剑影、炮声隆隆。

建筑总体造型的勾画取意自毛主席的书法——挥洒泼墨，着墨不多，却用寥寥几画一气呵成地塑造出深远意境之势。位于娄山关景区入口处的陈列馆既是一个独立的展馆，又是引导游客进入娄关山景区的必经之路。在总体造型中，我们以铜墙铁壁象征坚如磐石的雄伟关隘，以迂回转折的漫道将场地入口至上山步道的整个路径串联起来，象征行军之路的艰难险峻。娄山关属于中亚热带润湿季风气候，山区潮湿多雨，我们在场地的西南侧设置浅池收集雨水，将周边的苍山倒映在水中。水平如镜的水面恬静温柔，与周围若斧似戟的大山形成鲜明对比，并共同塑造出安宁而肃穆的纪念场所气氛。同时也形成取景框，将远山近水纳入画中，印证了毛主席当年"苍山如海"的壮丽诗句。

1　工具间 / Tools room
2　纪念品商店 / Souvenir store
3　出口门厅 / Exit lobby
4　屏幕控制 / Screen control
5　报告厅 / Lecture hall
6　下沉庭院 / Sunken courtyard
7　门厅 / Hall
8　序厅 / Lobby
9　展厅走道 / Passway of exhibition hall
10　沙盘区域 / Sand table
11　展厅 / Exhibition hall
12　疏散通道 / Evacuation exit
13　办公室 / Office
14　办公门厅 / Office hall
15　会议室 / Meeting room
16　升降平台 / Lift platform
17　文物修复室 / Office for restoration of cultural relic
18　消防控制室 / Fire control room
19　消防水池 / Fire pool
20　暖通设备室外平台 / Outdoor terrace for heating and ventilation facilities
21　地下车库 / Underground garage
22　卸货平台 / Terrace for unloading

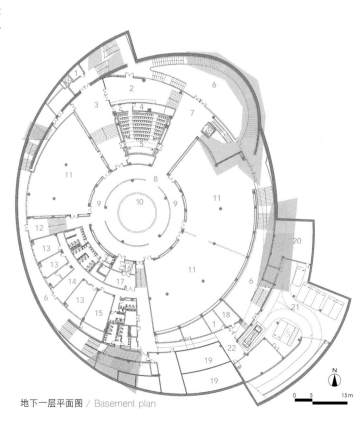

地下一层平面图 / Basement plan

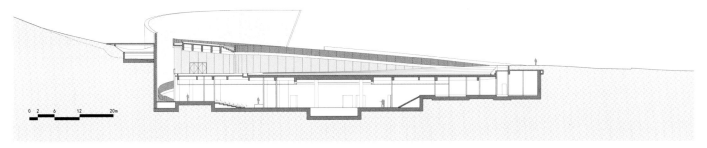

剖面图 / Section

02 主入口内部透视图 / Interior perspective of the main entrance
03 坡道内部透视图 / Interior perspective of the ramp

CHANGSHA INTERNATIONAL CONVENTION AND EXHIBITION CENTER

长沙国际会展中心

The Changsha International Convention and Exhibition Center has an area of 445,100 square meters covering the complex exhibition center, conference center, business center, shopping center, entertainment center, and reception center. Its main functions include convention and exhibition, conference, and ancillary facilities. It is a central venue within the surrounding functional area, and is one of the landmarks for Changsha's revitalization, further urbanization, and growing reputation.

总平面图 / Site plan

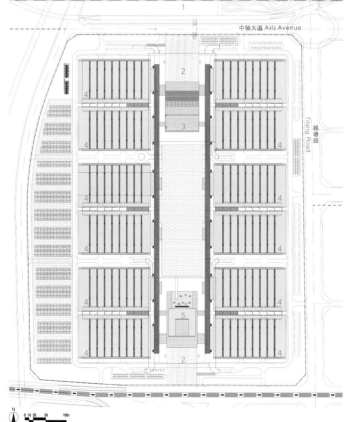

1 中轴公园 / Axis park
2 入口广场 / Entrance plaza
3 主登陆厅 / Main login hall
4 单层展馆 / Single-deck exhibition hall
5 次登陆厅 / Secondary login hall

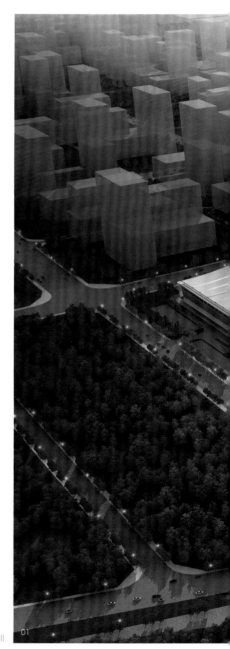

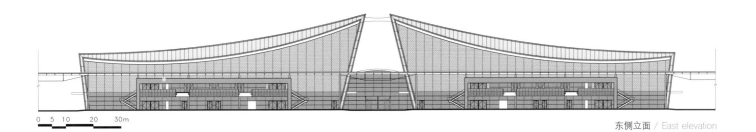

东侧立面 / East elevation

01 总体鸟瞰图 / Overall aerial view

长沙国际会展中心总用地面积 2026 亩。其中会展区 870 亩（含周边道路），配套区 1156 亩（含周边道路），中轴大道以南为会展展馆区，以北及区域南侧为会展配套区，考虑未来会展中心的发展，将梯塘路以东、中轴大道以南、金桂路以西规划为展馆的拓展区域。该用地西临浏阳河、紧邻武广高铁新城，位于城市空间结构中的南部综合发展带上，区位条件优越，是未来城市的副中心，与规划中的地铁 2 号线、4 号线相连，是城市重要的交通节点。

长沙国际会展中心楼体的总建筑面积为 44.51 万平方米，是集展览中心、会议中心、商务中心、购物中心、娱乐中心和接待中心于一体的复合功能区中的核心场馆，其主要功能包含会展、会议及相应配套设施，是长沙提升城市形象、完善城市功能、打造城市名片的标志之一。

长沙水文资源丰富，湘江横贯南北，整个城市临水而建，因水而兴，有"山水洲城"的美誉。会展中心紧邻湘江支流浏阳河，基地独特的地理位置，激发了设计灵感，设计撷取岳麓山山之意向，展馆沿河采用反弧形天际线，体现长沙的潇湘水韵，营造一幅浏阳河边的写意山水画。同时，沿河连续舒展的屋面，加强了从远处高铁站观看的标志性。

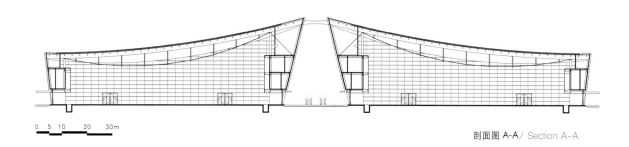

剖面图 A-A / Section A-A

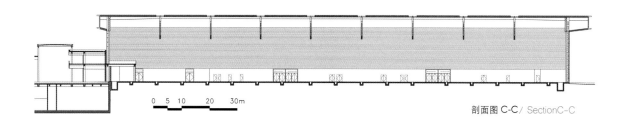

剖面图 C-C / SectionC-C

02 透视图（一）/ Perspective 1
03 透视图（二）/ Perspective 2
04 连廊透视图 / Perspective of the corridor
05 入口大厅 / Entrance hall

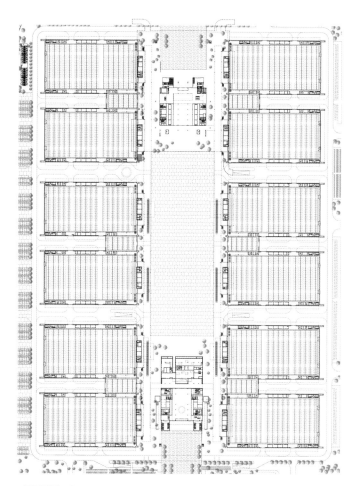

一层平面图 / First floor plan

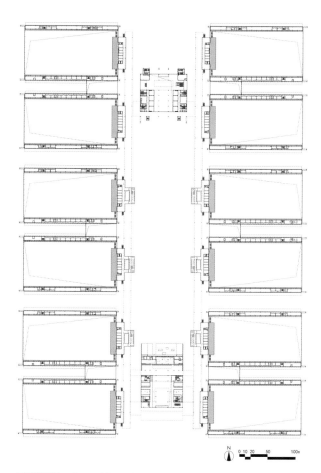

二层平面图 / Second floor plan

06 多功能厅效果图 / Rendering of the multi-function hall
07 中餐厅效果图 / Rendering of the Chinese dinning
08 登陆厅中庭 / Login hall courtyard
09 展厅内部透视图 / Interior perspective of the exhibition hall

SHANGHAI CHESS INSTITUTE
上海棋院

The Shanghai Chess Institute is located on Nanjing West Road, a prosperous commercial area of Shanghai. It measures 140 meters from south to north and 40 meters from east to west. The designer has arranged indoor and outdoor space to enclose a courtyard with walls and break the walls with the courtyard, squeezing every bit of available external space from the narrow space.

01 上海棋院 / Shanghai Chess Institute

总平面图 / Site plan

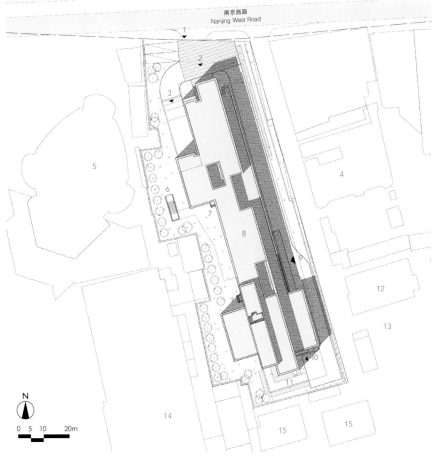

1 车行出入口 / Entrance for vehicles
2 比赛厅入口 / Entrance to play hall
3 地下车库出入口 / Entrance to underground garage
4 底商住宅 / Commercial and living building
5 广电大厦 / Broadcast & Television Edifice
6 人防出入口 / Entrance for air defense
7 疏散口 / Evacuation exit
8 上海棋院 / Shanghai Chess Institute
9 办公入口 / Entrance for staff
10 货运入口 / Entrance for freight
11 非机动车停车场 / Parking lots for non-motors
12 教学楼 / Teaching building
13 静安区社区学院南西分院 / Southwest Branch of Community College of Jing'an District
14 广电大厦演播楼 / Performance & Broadcast Building of Broadcast & Television Edifice
15 周边住宅 / Neighborhood residence

1 人防出口 / Exit for air defense
2 变电所 / Electrical substation
3 棋牌历史演示厅 / Presentation room for the history of chess and cards
4 门厅 / Hall
5 贵宾休息室 / VIP room
6 办公及贵宾入口 / Entrance for staff and VIP
7 200 席比赛大厅 / Play hall for 2000 seats
8 观众入口 / Entrance for audience
9 人行出入口 / Passage way

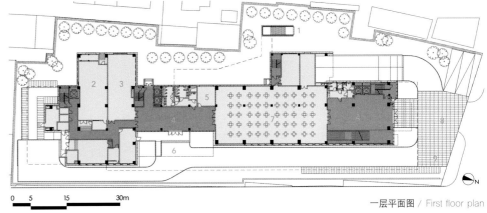

一层平面图 / First floor plan

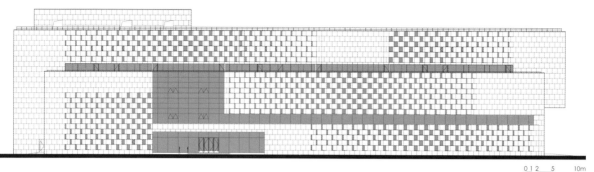

南侧立面图 / South elevation

上海棋院项目地处上海市繁华商业区南京西路，基地为南北向狭长地块，南北长约140米，东西最窄处约40米。

设计师将室内和室外的虚实空间交错布局，以墙围院，以院破墙，从而在狭小的用地内争取外部空间。

通过院与墙的结合，不仅融合了中国传统建筑的精髓，还以现代的手法体现传统空间，使建筑整体形态完整统一。而且庭院的运用使得建筑整体充满了中国意味。棋院以安静祥和的姿态出现在充满商业意味的南京西路，与周边建筑形成强烈的对比和反差，从而突出建筑的文化形象。

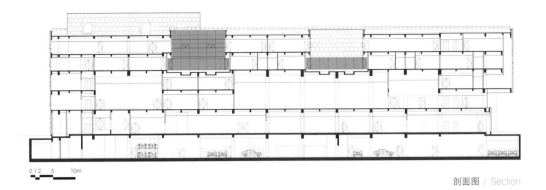

剖面图 / Section

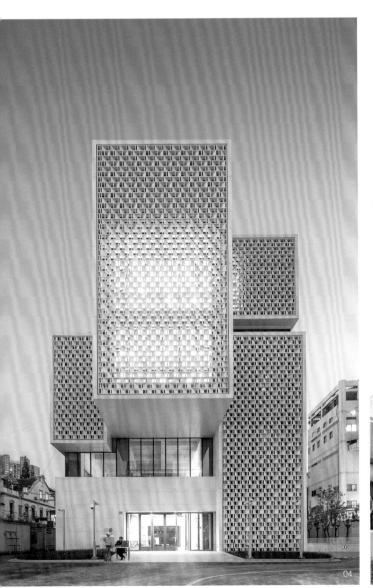

02 立面细节 / Façade details
03 办公入口 / Entrance for staff
04 比赛厅入口（一）/ Entrance to play hall 1
05 比赛厅入口（二）/ Entrance to play hall 2

WUZHONG DISTRICT DONGWU CULTURE CENTER
吴中区东吴文化中心

Located to the east of the Wuzhong Core Area's north-south axis, the Wuzhong District Dongwu Culture Center is located to the north of the Wuzhong District Government, thus enjoying an advantageous geographic location. It has four halls and two center buildings, including the Culture Hall, the Library, the Archive Hall, the Planning Exhibition Hall, the Conference Center and the Youth Activity Center.

总平面图 / Site plan

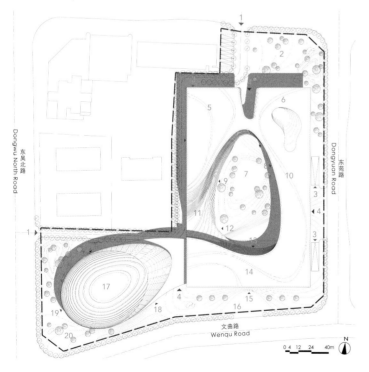

1 机动车出入口 / Entrance for vehicles
2 图书馆知识广场 / Knowledge Plaza of Library
3 地下车库出入口 / Entrance to underground garage
4 中央广场出入口 / Entrance to central plaza
5 档案馆，6楼 / Archive, 6F
6 图书馆，5楼 / Library, 5F
7 市民活动广场 / Civic Square
8 展厅次入口 / Secondary Entrance to Exhibition Hall
9 文化活动中心入口 / Entrance to Culture Hall
10 规划展示馆，4楼 / Planning Exhibition Hall
11 文化馆 / Cultural Hall
12 文化馆主入口 / Main entrance to Culture Hall
13 会议中心次入口 / Secondary entrance to Convention Center
14 会议中心，4楼 / Convention Center, 4F
15 会议中心主入口 / Main entrance to Convention Center
16 会议中心礼仪广场 / Etiquette square of Convention Center
17 青少年活动中心，5楼 / Youth Activity Center, 5F
18 青少年活动中心入口 / Entrance to Youth Activity Center
19 影城出入口 / Entrance to cinema
20 青少年广场 / Youth Plaza

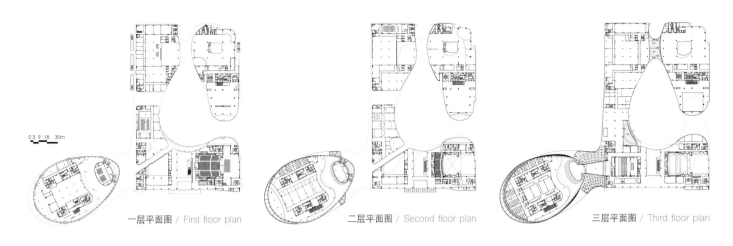

一层平面图 / First floor plan 二层平面图 / Second floor plan 三层平面图 / Third floor plan

01 鸟瞰图 / Aerial view

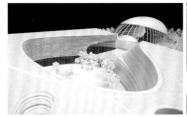   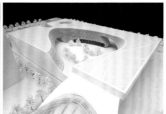

建筑模型 / Architectural models

吴中区东吴文化中心位于吴中区核心区南北中轴线东侧,吴中区政府北侧,地理位置优越。功能汇集四馆两中心,包括文化馆、图书馆、档案馆、规划展示馆、会议中心和青少年活动中心,是一座建筑综合体。

"万物之本,生命之水"——水是生命之源、吴中之源,亦是建筑设计概念之源。

"刚柔相成,万物乃形"——湖石经过水的精雕细琢,是展示吴中文化底蕴的最佳载体。建筑体量外刚内柔,汇集各类功能实体,通过阴阳结合的方式围合中央广场空间,形成一个容纳市民文化活动的容器。

"石水相生,钟灵毓秀"——形态饱满的水滴从湖石中悄然滑落,功能设置青少年活动中心,与水之活力相契合,彰显吴中文化的生命力。通过简洁抽象的建筑语言将其塑造为体现自由交流、多元并存的特色地标。

"随曲合方,宛若天开"——提炼江南园林空间精髓,追求自然之美,蕴藏诗画之意,内外空间融合,视线通透。

"智圆行方,秀外慧中"——建筑立面外刚内柔、流动连续,积极回应城市界面,保持城市公共空间的连续性。

02 内院 / Internal courtyard
03 南侧透视图 / Perspective of south elevation

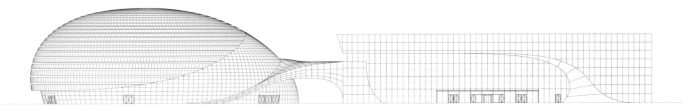

南侧立面图 / South elevation

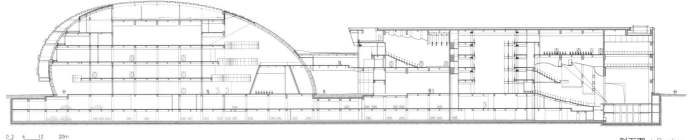

剖面图 / Section

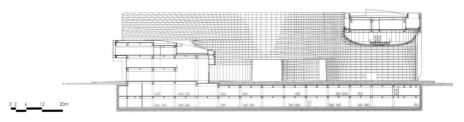

剖面图 / Section

FUZHOU URBAN DEVELOPMENT EXHIBITION HALL
福州城市发展展示馆

Located east of Fuzhou Nantai Island Strait International Convention and Exhibition Center, the base is a triangle with a wider south and narrower north, covering a floor area of 53,300 square meters. It neighbors the Puxia River to the south and faces the Minjiang River to the north. Based on the design concept of making the building both integral and transparent, unique yet harmonious, both classical and modern, this dignified museum is known as the "living room of Rongcheng."

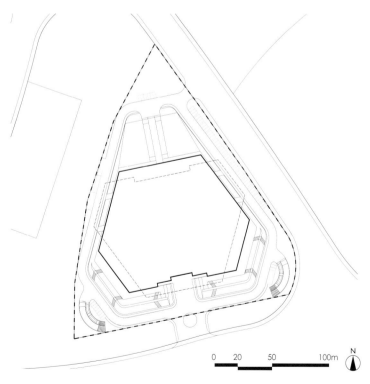

总平面图 / Site plan

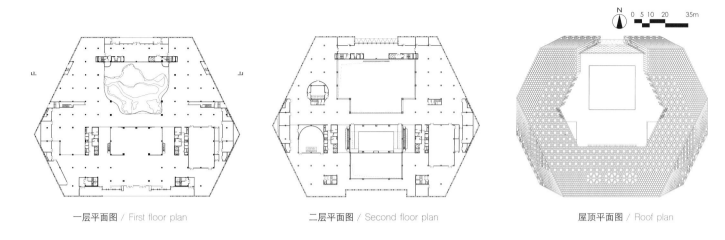

一层平面图 / First floor plan 二层平面图 / Second floor plan 屋顶平面图 / Roof plan

01 日景鸟瞰 / Aerial view

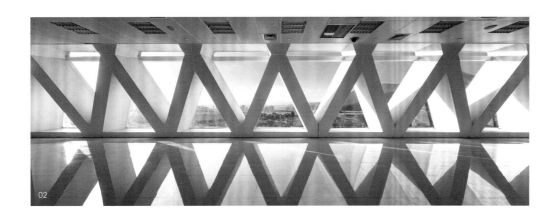

项目位于福州市南台岛海峡国际会展中心东侧，为南宽北窄的三角形，南依浦下河，北望闽江水。建筑面积53300平方米。建筑的钻石造型契合了基地南宽北窄的三角形轮廓，为争取更多的好朝向，避免拥塞感，建筑底层亦为南宽北窄的三角形，而向上至顶层则渐变为南窄北宽的倒三角形。本建筑将定位于既整体，又通透；既独特，又和谐；既有古典的端庄，又有现代的活力的"榕城客厅"。

福州又称"榕城"，自古已有"绿荫满城，暑不张盖"的景象。榕树的树干、根和枝丫交织在一起，从中可抽象出枝繁叶茂，独木成林的视觉效果，继而转化为福州市城市发展展示馆的立面肌理——由底层的透空柱廊向上逐渐演变为实墙面上的精巧窗洞。

福州特色之一的寿山石文化是以寿山石雕为载体。"寿山石"因其色彩斑斓、温润如玉、晶莹剔透，素有"石之君子""国之瑰宝"的美誉。展示馆的造型从中获取灵感，犹如屹立在闽江畔的文化巨石——晶莹剔透、端庄高雅。

福州因城内有屏山、乌山、于山，别称"三山"，因此展示馆体量的虚实布局契合了"三山一轴"的空间秩序，以示对古城文脉的尊重。

建筑的南立面退后城市道路约50米，向外倾斜，结合骑楼、敞廊形成了一个内凹的半室外的复合空间，提供市民等候、集散、休闲的场所。这既符合福州夏热冬暖的气候特征，又促进了建筑内外空间的相互渗透，增强了人流的导向性。

02 **室内局部** / Part of interior view
03 **立面局部** / Part of the façade
04 **日景透视** / Perspective

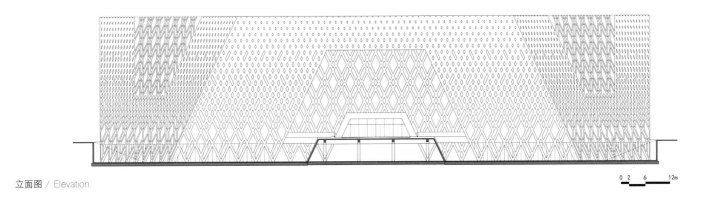

立面图 / Elevation

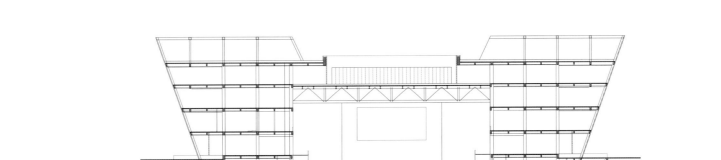

剖面图 / Section

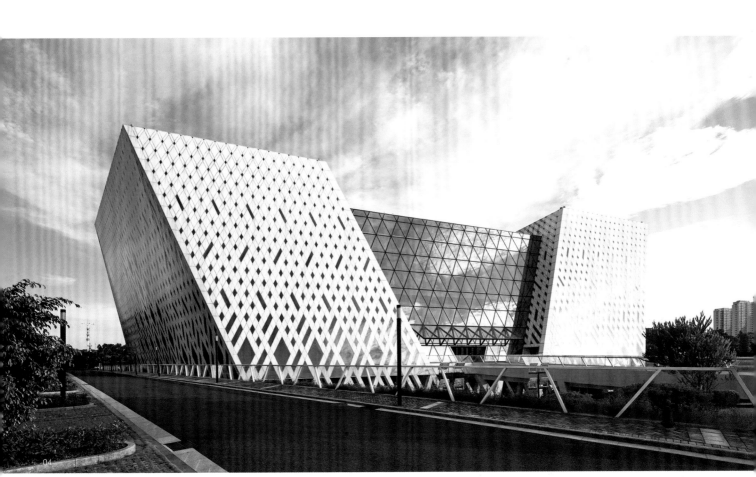

RELOCATION OF THE SHANGHAI SYMPHONY ORCHESTRA

上海交响乐团迁建工程

Located north of Fuxing Middle Road, and east of Baoqing Road, it is the former site of the Shanghai Diving Pool. Built on an area of 16,318 square meters with a gross floor area of 19,950 square meters, it has 5,274 square meters above ground and 14,676 square meters underground. Metro Line 10 runs across the southeast corner of the base along Fuxing Middle Road, and this creates some audio disturbance to the building.

总平面图 / Site plan

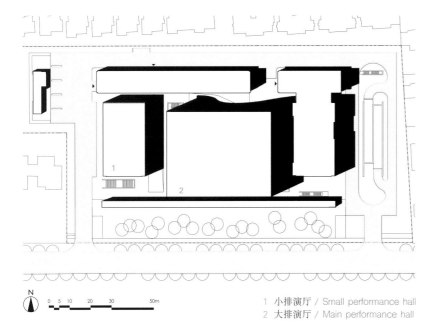

1 小排演厅 / Small performance hall
2 大排演厅 / Main performance hall

01 总体透视图 / Overall perspective view
02 鸟瞰图 / Aerial view

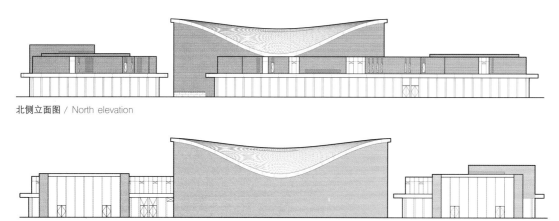

北侧立面图 / North elevation

南侧立面图 / South elevation

项目位于复兴中路北侧、宝庆路东侧，原上海跳水池所在地。用地面积 16318 平方米。总建筑面积 19950 平方米，其中地上建筑面积 5274 平方米，地下建筑面积 14676 平方米。地上两层、地下四层，其中地上及地下一、二层为使用层，地下三、四层为设备管道及隔振设施专用空间及设备机房。

该地区位于上海衡山复兴历史风貌保护区，衡山路—复兴路历史文化风貌区是中国上海市立法保护的历史文化风貌区之一，也是中心城区 12 个历史文化风貌区中规模最大的一个。交响乐团的建筑体量和高度的确定除了尊重城市规划指标要求外，还充分考虑了建筑对周边居民的影响，尽可能地将大且相对高的体量设置在远离居民小区的道路一侧，并逐渐向北面居民小区一侧递减高度，使建筑对居民日照、景观、天际线及开挖对其的影响降低到最小。由于音乐厅必须由室内空间容量及高度来保证音响效果的特殊性质，所以 1200 座大排演厅的高度向地上及地下两个方向扩展。由于地下开挖深度受经济、基础结构、建筑整体动线、地铁等多方面的因素制约，大排演厅的地上高度局部达到 18 米。为降低建筑高度所带来的体量压迫感，大排演厅采用了曲面屋顶，降低了地面建筑高度。

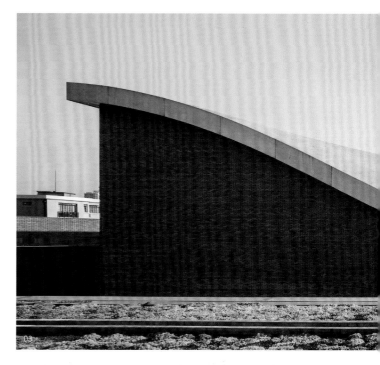

为了使本工程达到世界一流的音响水准，设计师采用了"套中套"的结构形式和隔而固避振弹簧来降低地铁 10 号线对排演厅的噪音干扰。排演厅 A、B 自身采用混凝土双层墙、顶、底板结构，形成独立密闭的空间，以隔绝所有空气传声；两个排演厅与建筑其他部位完全断开，以隔绝任何固体传声的可能；在两个排演厅下方的结构大底板上设置隔而固隔振弹簧承托两个排演厅结构，以彻底隔绝来自地铁 10 号线的振动影响。

03 大厅曲面屋顶 / Curved roof of the hall
04 内庭院夜景 / Night view of internal courtyard

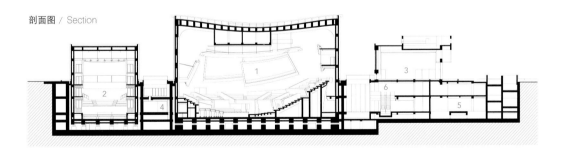
剖面图 / Section

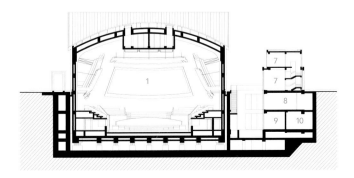

1 大排演厅 / Main performance hall
2 小排演厅 / Small performance hall
3 入口大厅 / Entrance hall
4 演员候场大厅 / Waiting hall for performers
5 地下停车场 / Underground parking
6 观众候场大厅 / Waiting hall for audience
7 办公室 / Office
8 休息厅兼展区 / Resting hall & exhibition area
9 演员休息厅 / Resting room for performers
10 排练室 / Rehearse room

05 门厅 / Hall
06 大厅室内（一）/ Interior view of the hall 1
07 大厅室内（二）/ Interior view of the hall 2
08 大厅室内（三）/ Interior view of the hall 3

一层平面图 / First floor plan

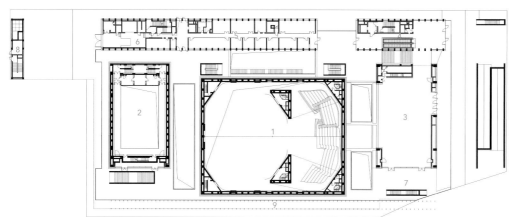

1 大排演厅 / Main performance hall
2 小排演厅 / Small performance hall
3 入口大厅 / Entrance hall
4 VIP入口大厅 / Entrance hall for VIP
5 办公室 / Office
6 卸货区 / Unloading area
7 室外演艺广场 / Outdoor performance plaza
8 设备用房 / Facilities room
9 绿茵走廊 / Green corridor

JIADING NEW TOWN POLY GRAND THEATER

嘉定新城保利大剧院

Located in Jiading District, Shanghai, Plot D10-15 of Jiading New Town neighbors Yuanxiang Lake to the southeast. With a total area of 30,235 square meters and a gross floor area of 55,904 square meters, the Poly Grand Theater has two halls, including 1,543 seats in the grand hall and 498 seats in the multi-functional hall. It is a large-scale theater.

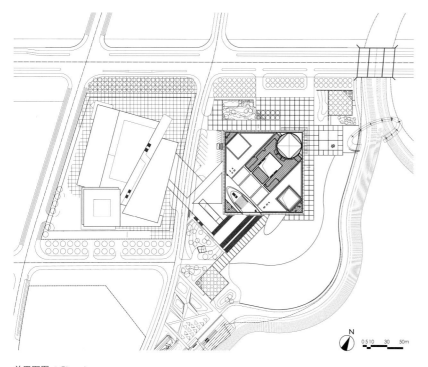

总平面图 / Site plan

01 东南角水面夜景 / Night view of the waterscape from southeast
02 万花筒内景 / Interior view of the kaleidoscope
03 东南立面夜景 / Night view of southeast façade

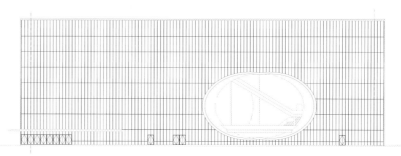

北侧立面图 / North elevation

项目位于上海市嘉定区，嘉定新城 D10-15 地块，东南方向面临远香湖。总用地面积 30235 平方米，总建筑面积 55904 平方米。大剧院共有两个厅，其中大观众厅 1543 座，多功能厅 498 座，为大型剧院。

设计的主要构思是将剧院设计成为"文化的万花筒"。如同将周围的光线导入，通过漫反射展现如万花筒一般的光影效果。

在形态操作上，大剧场以一个 100 米 ×100 米 ×34 米的立方体形式展开，在基地中构成了中心。所谓"万花筒"是通过 5 组直径 18 米的圆筒以不同的方向与立方体相交，在保证核心观众厅功能的基础上，将光、水、风等自然要素以及周边远香湖的自然美景有机的引入到建筑内部，从而在简洁型体的内部形成了丰富变化的室内和半室外的公共空间。在这里聚集的人们可以同自然和风景对话，同不同年龄、不同背景的人对话，在这里，空间成为文化交流的场所。

在外立面设计上，单片超白玻璃幕墙设置在了清水混凝土外侧，还在底部和顶部开设了百叶，通过内部气流的有效组织，使形成的空腔达到双层幕墙的绿色节能效果。轻盈的玻璃表皮通过水面的倒影，有效的缓解了大体量清水混凝土立面的压迫感，同时通过泛光照明的衬托，使得剧院建筑的文化特质得到有力的体现。

在建筑材料的设计上，由外到内依次通过玻璃、清水混凝土、外饰木纹的铝合金和木材的过渡，由冷转暖。通过材质的变化给人以身心愉悦之感。

在景观设计上，在紧邻建筑的南侧及东侧设计了连接公园区域的水池，使得大剧院和湖景自然地融为一体。水边的广场与剧院北侧的前广场相连，形成可以让市民休闲体验的漫步道。

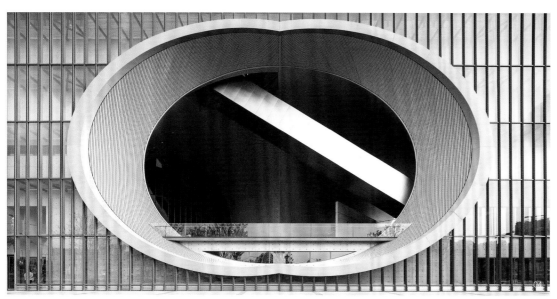

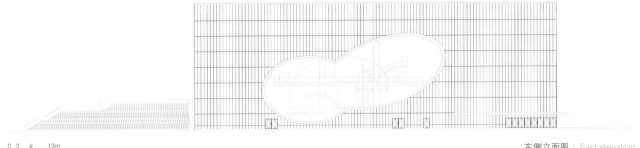

东侧立面图 / East elevation

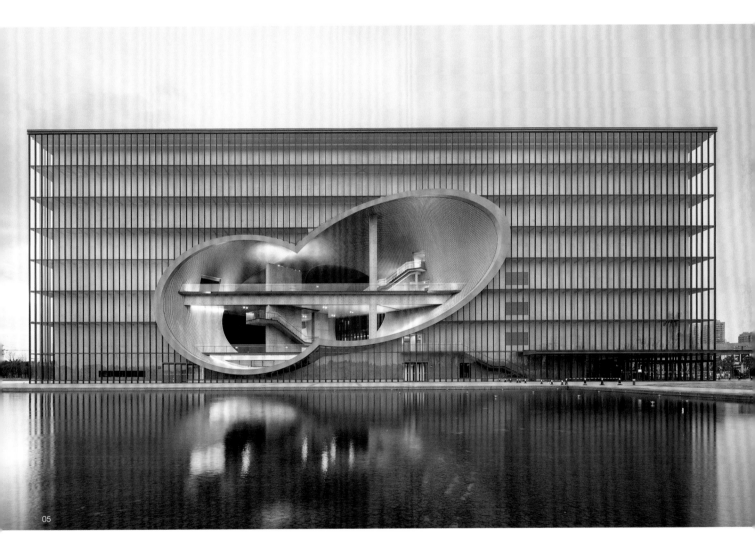

04 北立面细部 / Details of the north façade
05 东立面夜景 / Night view of the east façade

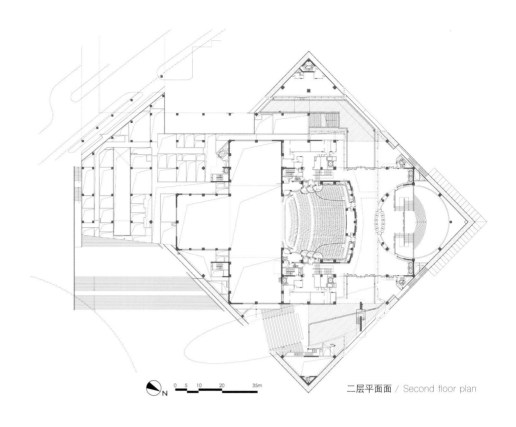

二层平面面 / Second floor plan

06 观众厅序厅 / Auditorium lobby
07 观众厅看台 / Grandstand of the auditorium

WANDA WUHAN MOVIE PARK
武汉电影乐园

Located in the cultural tourism area of Wuhan's Central Culture District, Wanda Wuhan Movie Park has gathered the world's most cutting-edge movie and entertainment technology. It has six themed cinemas, namely, a 4D cinema, a 5D cinema, a flying cinema, an interactive cinema, a spatial cinema, an experiential cinema, as well as themed shopping, dining, and other service facilities. It's a complex offering entertainment, business, and dining.

1 沙湖公园 / Shahu Park
2 货运出入口 / Entrance for freight
3 绿地 / Green land
4 地下车库出入口 / Entrance to underground garage
5 后勤办公出入口 / Entrance for staff
6 武汉电影乐园 / Wuhan Movie Park
7 主入口雨棚 / Canopy of main entrance
8 建筑主入口 / Main entrance
9 步行广场 / Pedestrian plaza
10 商店出口 / Shop exit
11 机动车出入口 / Entrance for vehicles

总平面图 / Site plan

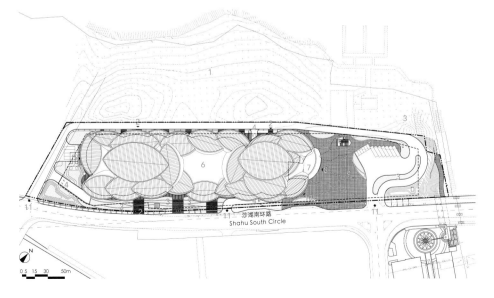

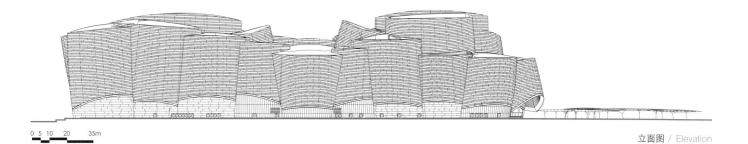

0 5 10 20 35m

立面图 / Elevation

01 对岸远景 / Distant view from the opposite bank

115

武汉电影乐园位于武汉中央文化区文化旅游区。作为全球首个室内电影文化公园，汇集全球最新顶尖电影娱乐科技，容纳六大娱乐主题——包括4D影院、5D影院、飞行影院、互动影院、时空影院、体验影院，以及电影主题购物、餐饮等服务设施。是以电影体验为主，集娱乐、商业、餐饮于一体的建筑综合体。

主体建筑造型概念来自于楚汉文化精髓——"编钟"。编钟形态偎依交错，形成两组体量，尺度上满足分层容纳大型娱乐设施的要求；两组体量之间形成中庭空间，利用采光天窗将自然光引入室内。建筑首层立面采用两个楼层高度的玻璃幕墙，令建筑室内拥有朝向沙湖公园的良好视野，同时也为内部餐饮、购物等功能提供明亮通透的商业氛围。建筑上部立面"编钟"采用金色的铝合金幕墙，复合亮面和哑光材质表面，以不同的方式捕捉和反射光线。夜晚，整个建筑物的夜景照明烘托了外立面幕墙暖金色的色调，每隔4米左右的水平向LED勾边光带沿着整个外立面延展，将"编钟"的造型描绘在城市夜空中。

设计以绿色建筑二星级要求为指导，应用诸多节能技术，确保建筑全寿命周期内最大限度地节地、节能、节水、节材、环保，实现可持续发展。

02 鸟瞰图 / Aerial view
03 主入口 / Main entrance
04 近景 / Close shot

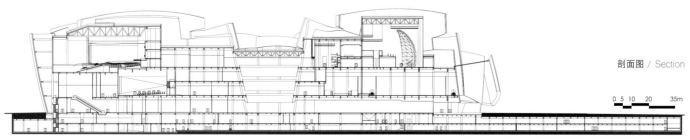

剖面图 / Section

0 5 10   20   35m

一层平面图 / First floor plan

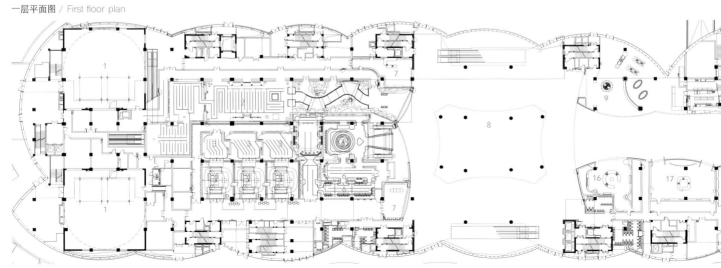

05 小中庭 / Small atrium
06 主中庭 / Main atrium
07 4D 影院入口 / 4D cinema entrance

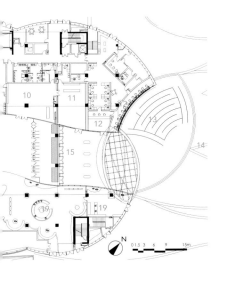

1   飞行影院 / Flying cinema
2   飞行影院排队区 / Queue area to flying cinema
3   预演厅 / Rehearsal hall
4   时空影院 / Spacetime cinema
5   飞行影院观影通道 / Passageway to flying cinema
6   时空影院排队区 / Queue area to spatial cinema
7   出口商店 / Shop at the exit
8   中庭 / Courtyard
9   餐厅 / Dining hall
10 VIP 休息室 / VIP lounge
11 VIP 接待室 / VIP reception room
12 售票处 / Ticket office
13 排队区 / Queue area
14 建筑主入口 / Main entrance to building
15 无障碍入口 / Barrier-free entrance
16 游乐项目摄影零售 / Recreation, shooting and retailing
17 商店 / Shops
18 便利店 / Convenience store
19 公园出口商店 / Shop at the exit

PLOT NO. 10, HAINAN ROAD, HONGKOU DISTRICT, SHANGHAI

上海市虹口区海南路 10 号地块项目

Located on Wujin Road and Wusong Road in the Hongkou District of Shanghai, there is a building under protection on the southwest corner, which faces municipal roads on three sides, and has the Metro Line 10 running through it from east to west. It's a plot for offices and supporting commercial facilities, with an area of 16,427 square meters and a gross floor area of 95,077 square meters.

01 外观日景 / Exterior view

总平面图 / Site plan

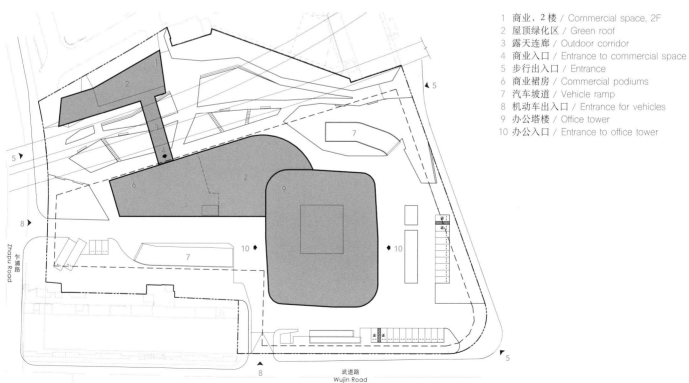

1 商业，2 楼 / Commercial space, 2F
2 屋顶绿化区 / Green roof
3 露天连廊 / Outdoor corridor
4 商业入口 / Entrance to commercial space
5 步行出入口 / Entrance
6 商业裙房 / Commercial podiums
7 汽车坡道 / Vehicle ramp
8 机动车出入口 / Entrance for vehicles
9 办公塔楼 / Office tower
10 办公入口 / Entrance to office tower

02 入口外观 / Entrance
03 立面细部 / Details of building façade

项目位于上海市虹口区吴淞路武进路，西南角有一幢保护建筑，三边临城市道路，地铁十号线自西向东穿越本地块。项目功能为办公及配套商业，建筑用地面积16427平方米，总建筑面积95077平方米。

整组建筑采用独特的垂直感外遮阳百叶。以18毫米宽的白色铝条编织成具有通透感，犹如蕾丝网格般的百叶，给予建筑不断的明暗变化。根据阳光的角度、强度及颜色的变换，建筑立面的纹理也随之产生不同的光影变化——时而锐利，时而柔和。这一弯折百叶系统在平面上呈三角形拼接，在高度上逐渐变化，形成有韵律的构件，借此创造出与一贯以坚硬冰冷素材所组成的高楼所截然不同的，更为柔性的建筑。

项目获得美国绿色建筑LEED金奖认证，为四川北路商圈提供了高品质的办公空间。建筑立面采用全玻璃幕墙体系，玻璃使用中空夹胶玻璃，在满足建筑节能要求的同时，也保证了建筑的安全性。建筑采用金属外遮阳系统。由白色铝条编织成的透空遮阳百叶，在不影响室内视野的情况下，使建筑获得很好的遮阳效果，并将玻璃幕墙的反射光对周边环境的影响降到最低。在裙房上设置屋顶绿化，减少顶层空调能耗，降低城市热岛效应，也使办公塔楼得到更好的视觉外观。

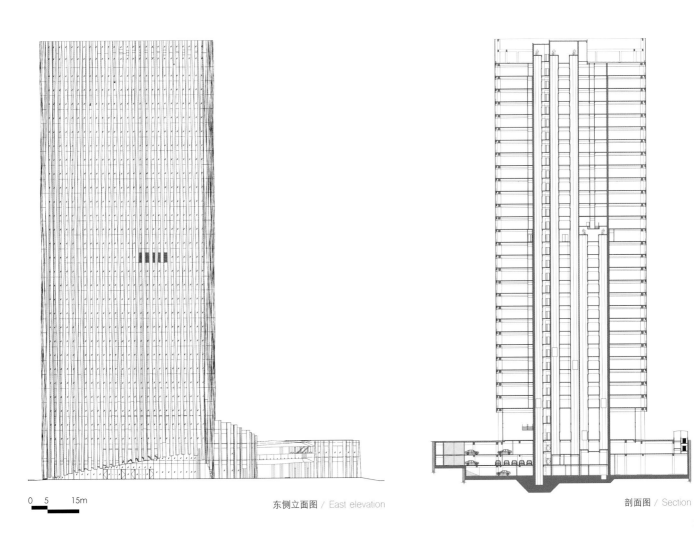

0 5 15m

东侧立面图 / East elevation

剖面图 / Section

123

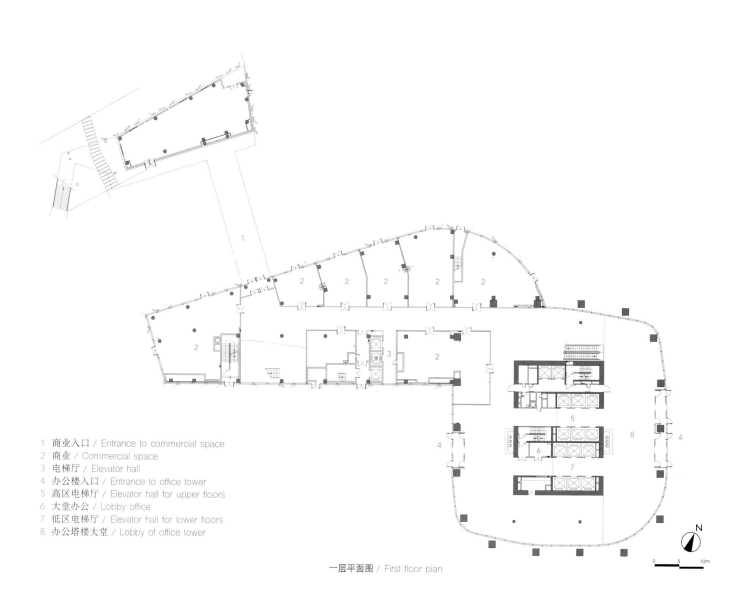

1 商业入口 / Entrance to commercial space
2 商业 / Commercial space
3 电梯厅 / Elevator hall
4 办公楼入口 / Entrance to office tower
5 高区电梯厅 / Elevator hall for upper floors
6 大堂办公 / Lobby office
7 低区电梯厅 / Elevator hall for lower floors
8 办公塔楼大堂 / Lobby of office tower

一层平面图 / First floor plan

04 室内空间（一）/ Interior space 1
05 室内空间（二）/ Interior space 2
06 室内空间（三）/ Interior space 3
07 内部大堂 / Internal lobby

OFFICE BUILDING OF SHANGHAI CITY CONSTRUCTION AND INVESTMENT CORPORATION

上海市城市建设投资开发总公司企业自用办公楼

Located within the ecological area of New Jiangwan Town, the office building of Shanghai City Construction enjoys convenient transportation and a charming environment. With a gross floor area of 21,968 square meters, the design has fully adapted the building to and regulated its impact on the surrounding environment. It has been granted with 3-Star Green Building (issued by MOHURD) design award thanks to its energy-efficient technology and design.

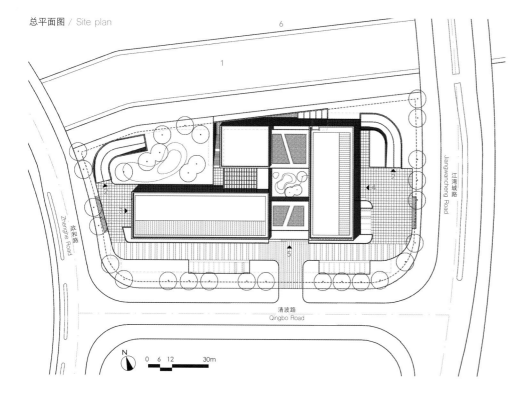

总平面图 / Site plan

1 水体 / Water body
2 地下车库出入口 / Entrance to underground garage
3 会议中心 VIP 入口 / VIP entrance to Convention Center
4 科技研发中心入口（B 栋）/ Entrance to Technical Research and Development Center (Building B)
5 主入口 / Main entrance

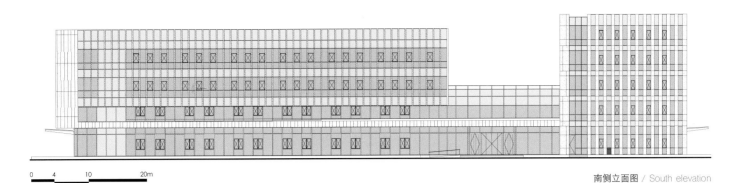

南侧立面图 / South elevation

01　沿河立面 / Façade view along the riverside

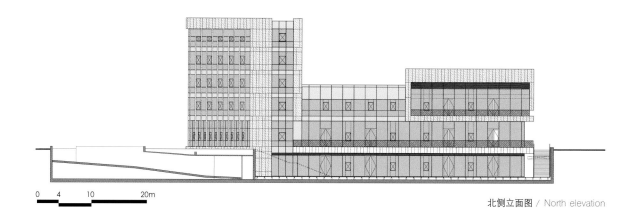

北侧立面图 / North elevation

上海城投企业自用办公楼位于新江湾城生态区内，交通便捷，环境优良。总建筑面积为21968平方米。

新江湾城优异的自然景观条件在本项目中与建筑有机结合。在建筑外部，场地北侧的河流与设计场地紧密联系，退让的绿色景观带和水体自然的融合在一起。在建筑内部，外部景观与内部空间自然渗透融合，自然的气息通过体量中围合的绿化中庭和北侧的绿化边庭以及屋顶绿化融入到各层的办公空间，达成了建筑与自然和谐共生的设计理念。

本设计充分考虑了建筑自身对环境的适应和调节，通过简洁规整的建筑体量，对自然采光及自然通风的充分利用以及合理的窗墙面积比和外遮阳的设置，减少了对机电系统的依赖，降低资源消耗。通过各项节能技术和设计的运用，取得了三星级绿色建筑设计标识。

L形的平面布局赋予建筑最大的景观接触面，同时通过中间的入口大厅将建筑中相互独立的两部分很自然的联系起来，高度不同的三块体量在建筑中心围合成一个内院，相互之间联系便捷高效。L形的建筑平面与北面的河流界定出一个面向河流开敞的庭院，将景观很好的引入建筑内部。

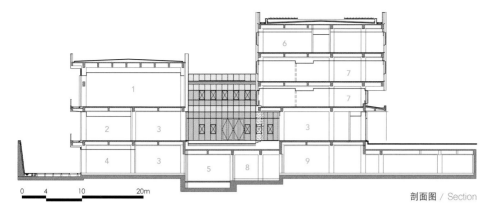

1 报告厅 / Lecture hall
2 健身房 / Fitness room
3 空调机房 / Air conditioning facilities room
4 VIP 餐厅 / VIP dining
5 变电所 / Electrical substation
6 管理人员办公室 / Office for administrative staff
7 敞开式办公室 / Open office
8 冷冻机房 / Refrigerator room
9 地下停车库 / Underground garage

剖面图 / Section

02 沿城市道路主立面 / Main façade along the street
03 下沉庭院 / Sunken courtyard
04 办公楼主入口 / Main entrance to office building

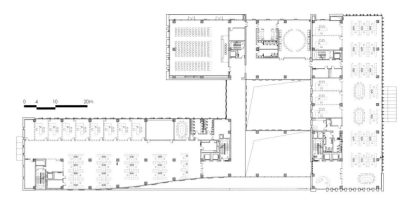

二层平面图 / Second floor plan

05 主入口门厅 / Main entrance hall
06 绿化中庭 / Green courtyard

一层平面图 / First floor plan

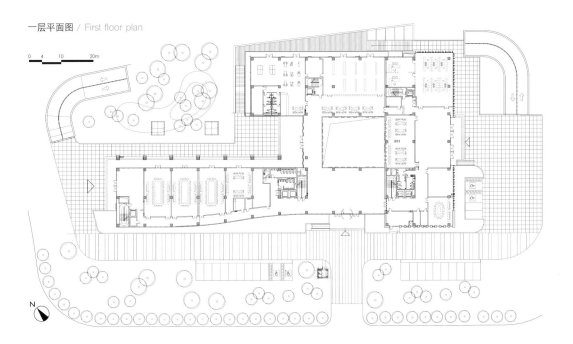

PROJECT OF NO. 22 PINGLIANG STREET

平凉街道 22 街坊项目

Neighboring Changyang Road to the north, Huoshan Road to the south, Jingzhou Road to the west and Liaoyang Road to the east, this project is a core component of the Dalian Road Headquarters R&D Cluster. Positioned as a large urban complex, the project will be built into premium office buildings and high-quality business facilities. With an area of 33,000 square meters and a gross floor area of 208,000 square meters, the project is composed of four high-rise office buildings, five multi-floor commercial buildings, and a three-layer underground garage.

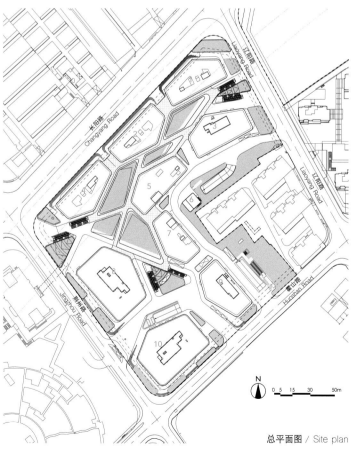

总平面图 / Site plan

1　H 楼（商业）/ Building H (Commercial use)
2　G 楼（商业）/ Building G (Commercial use)
3　J 楼（商业、办公）/ Building J (Commercial use & office)
4　F 楼（商业）/ Building F (Commercial use)
5　E 楼（商业）/ Building E (Commercial use)
6　P 楼（垃圾房）/ Building P (Refuse room)
7　D 楼（商业）/ Building D (Commercial use)
8　A 楼（商业、办公）/ Building A (Commercial use & office)
9　C 楼（商业、办公）/ Building C (Commercial use & office)
10　B 楼（商业、办公）/ Building B (Commercial use & office)

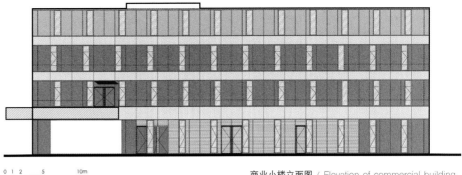

商业小楼立面图 / Elevation of commercial building

01　效果图 / Rendering

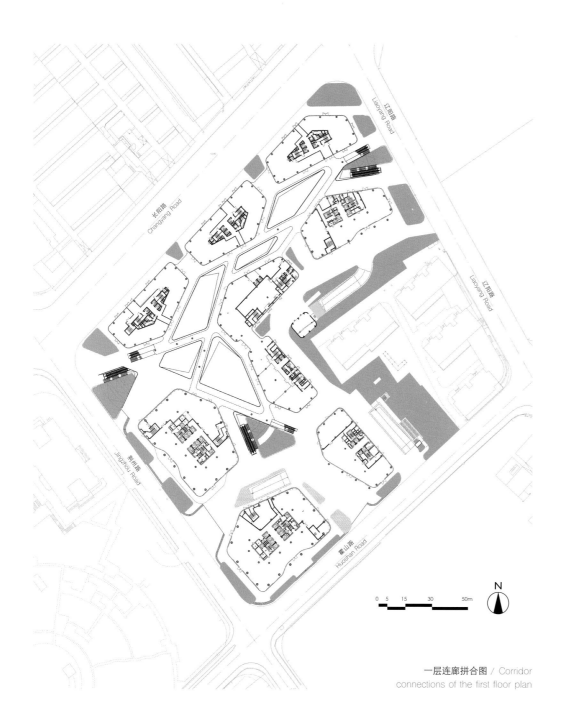

一层连廊拼合图 / Corridor connections of the first floor plan

项目北临长阳路、南临霍山路、西临荆州路、东临辽阳路，是大连路总部研发集聚区的核心组成地块。该项目规划设计定位为大型城市综合体，集甲级办公楼、精品商业设施于一体。建筑用地面积 3.3 万平方米，总建筑面积 20.8 万平方米，主要由四幢高层办公楼、五幢多层商业空间和三层的地下车库组成。

项目规划设计为一个大型城市综合体，连接两个相邻的地块，设计思路主要基于两个地块之间的连接、与用地西侧现状公园的连接，以及依据相邻交通流线形成的多通道现状。

建筑采用清晰的基本几何造型，以五边形为基础。虽然建筑综合体的整体外观具备生动而富有趣味的特色，但整个系统却是建立在一个模块化的体系上，仅通过旋转或镜像反射对各种造型进行组合，所有建筑共同构成了一个协调的建筑综合体，拥有独一无二的特征。

独立的圆角建筑引导着综合体内的行人；商业空间精品内街沿线得以最大限度的拓展。圆角的建筑体量形成于、并顺应于流畅的人流动线，展现出有机的入口姿态。协调的流线在建筑内部随着流动的桥梁继续伸展，将所有的商业空间相互连接起来，从而营造了一个舒适、轻快、开放的购物氛围。

02 室内效果图 / Interior rendering
03 连廊效果图 / Corridor rendering

BUILDING A OF TONGJI TOWER

同济大厦 A 楼

Located at the intersection of Siping Road and Zhangwu Road, Building A of Tongji Tower is opposite to the entrance of Tongji University. Standing at one end of the axis of Tongji University, the building has a floor area of 36,982 square meters which accommodates both offices and classrooms.

01　入口局部 / Entrance to the building
02　西北侧外观 / Northwest façade

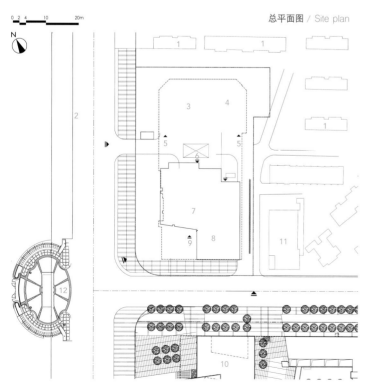

总平面图 / Site plan

1　同济新村住宅 / Residence of Tongji New Village
2　非机动车停车场 / Parking lots for non-motors
3　地面植草砖停车场 / Parking lots paved with grass-planting tiles
4　人防疏散楼梯 / Civil air defense evacuation stair
5　地下车库入口 / Entrance to underground garage
6　办公楼日常出入口 / Entrance to office building
7　同济大厦 A 楼 / Tongji Tower, Building A
8　屋面绿化 / Green roof
9　办公楼人行出入口 / User's entrance to office building
10　同济联合广场（南区）/ Tongji Union Square (Southern campus)
11　商业建筑 / Commercial buildings
12　同济大学校门 / Gate of Tongji University

同济大厦A楼位于上海市四平路和彰武路口，正对同济校门，位于同济大学校园主轴线的一端，建筑面积36982平方米，主要功能是办公和教学。

建筑主体平面采用了梯形，与南侧同济联合广场主楼共同围合出一个外向的公共空间。斜边的运用弱化了空间由开放、宽阔的校门向狭窄的线性道路的收缩，增强了整个建筑群对同济大学校门的导向性，进一步延伸了同济大学的主轴线。

建筑立面采用玻璃幕墙和石材结合的做法，利用玻璃和石材的错位组合使得立面和谐统一而具有韵律感，从而避免了大面统一幕墙的单调和呆板。由于建筑的西面正对同济大学校园，它既是可以鸟瞰同济校园的极好的观赏点，同时又极大地影响着校园本身的景观，因此建筑的西面隔层设置了中庭，这既丰富了建筑的室内空间，又为人们提供了一个休息并观赏同济校园风光场所。在中庭处采用双层幕墙，丰富了校园景观的同时也兼顾了建筑的西面遮阳。

对建筑底层的开放性设计是本方案的另一个特点，突出主体建筑的裙房底层大面积架空，与地铁10号线同济大学站出入口统一设计，并以此将室外的公共广场、半室外的地铁出入口和室内的开放空间联系起来，进而加强了建筑底部的通透感和开放性，突出了城市、校园与建筑的交汇与融合。

03　二层室内局部 / Interior view of the second floor
04　屋顶花园 / Roof garden

03

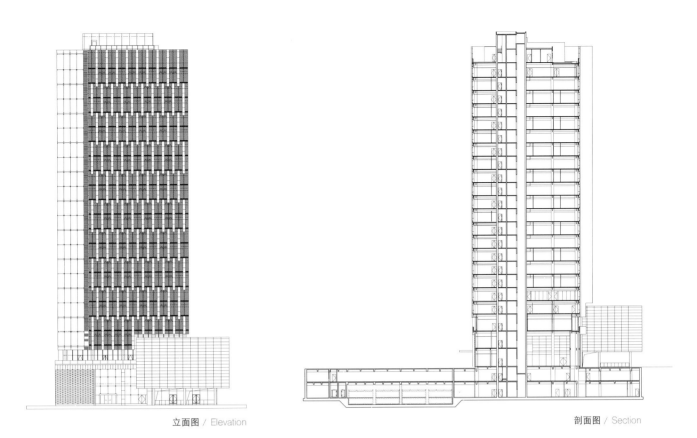

立面图 / Elevation

剖面图 / Section

04

139

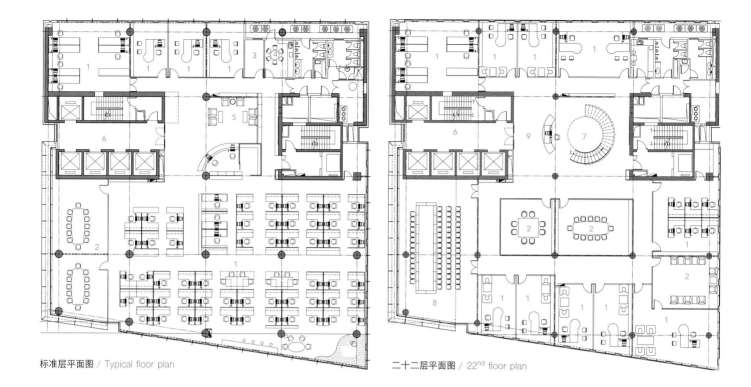

标准层平面图 / Typical floor plan

二十二层平面图 / 22nd floor plan

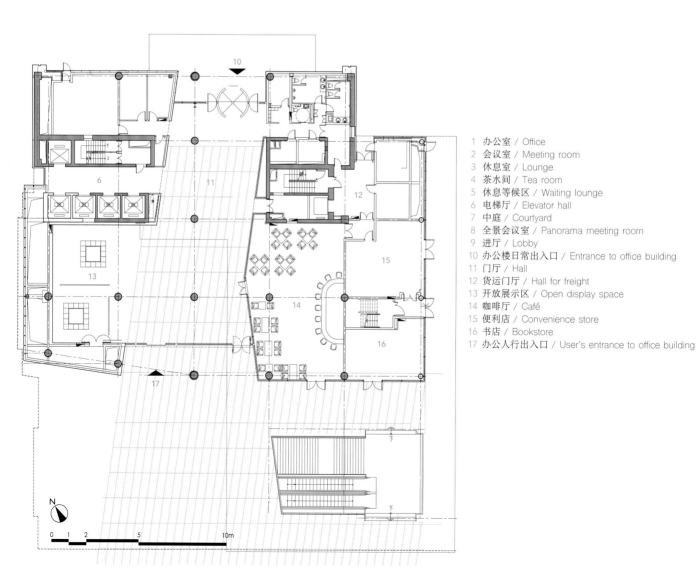

1 办公室 / Office
2 会议室 / Meeting room
3 休息室 / Lounge
4 茶水间 / Tea room
5 休息等候区 / Waiting lounge
6 电梯厅 / Elevator hall
7 中庭 / Courtyard
8 全景会议室 / Panorama meeting room
9 进厅 / Lobby
10 办公楼日常出入口 / Entrance to office building
11 门厅 / Hall
12 货运门厅 / Hall for freight
13 开放展示区 / Open display space
14 咖啡厅 / Café
15 便利店 / Convenience store
16 书店 / Bookstore
17 办公人行出入口 / User's entrance to office building

一层平面 / First floor plan

05 会议厅室内 / Interior of conference hall
06 旋转楼梯 / Spiral staircase

PRODUCTION BASE OF THE 32ND RESEARCH INSTITUTE OF THE CHINA ELECTRONICS TECHNOLOGY GROUP (JIADING)

中国电子科技集团第三十二研究所科研生产基地（嘉定）

Located in Jiading District, Shanghai, the Production Base of the 32nd Research Institute of China Electronics Technology Group (Jiading) has a total area of 19.08 hectares and a gross floor area of 280,000 square meters. The Phase I of the project consists of offices and R&D buildings, the employee cafeteria, an activity center, a quality inspection building and support buildings, including a comprehensive R&D park with office, scientific research facilities, production and quality inspection buildings, and dining and living facilities.

总平面图 / Site plan

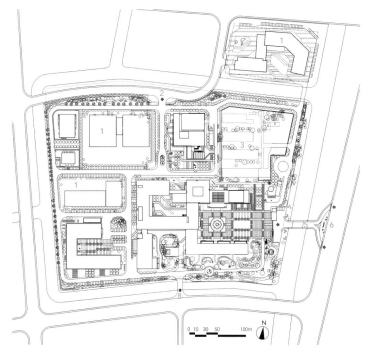

1 三期用地 / Land for Phase 3
2 北次要出入口 / North secondary entrance
3 整体保留生产车间 / Reserved workshops
4 保护植物区 / Protective plants
5 东门桥 / Bridge to east gate
6 东主要出入口 / East main entrance
7 景观小广场 / Landscape square
8 南次要出入口 / South secondary entrance

01 东南角鸟瞰图 / Aerial view from southeast
02 员工餐厅与活动中心 / Dining and activity center for staff
03 办公研发楼，西南庭院人视图 / R&D office building, view from southwest courtyard

中国电子科技集团第三十二研究所科研生产基地（嘉定）位于上海市嘉定区，园区总用地面积19.08公顷（含代征河道及保护绿地），总建筑面积28万平方米，其中一期工程由办公科研楼、员工餐厅与活动中心、质检大楼及园区配套用房组成。地上最高21层，地下1层。该基地是兼具办公、科研、生产、质检、餐饮及活动配套等服务的综合性办公研发园区。

项目选址于原三十二所老所区基地上的综合性办公园区，设计中将"自然和历史"这两大主题纳入到整体设计当中。

沉淀了六十年的老所区环境是三十二所基地的特点，一方面是林木繁茂的自然景观（河道、树木），一方面是有着特殊时代记忆的人文景观（毛主席广场）。这两大资源为有着六十多年历史的老所区披上了美丽而神秘的面纱，是非常难得的景观资源。

在园区的总体布局中，除了满足分区合理、交通便捷等理性因素，有意识的以现状树木作为设计的切入点，以环境作为空间的核心，通过建筑的围合、避让，最大化的保留了园区现有的树木、历史景观小品等，创造富有特点的园区环境。

园区的高容积率，使我们希望通过让建筑往空中发展，留出更多的地面空间，让建筑与环境更多地融合，创造出有品质的建筑空间。在建筑设计中，通过将各栋建筑分解、叠加、融合，以"院落围合"的空间形式将各部分有机串接起来，大大提升了建筑本身的使用效率，形成丰富而有趣的空间层次。

建筑造型以简洁的现代建筑语言，通过将建筑体量的上下两段切割、组合，在视觉和感观上消解大体量建筑可能形成的压抑感，也使得整个园区灵动、活跃。

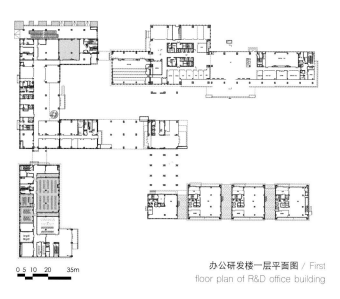

办公研发楼一层平面图 / First floor plan of R&D office building

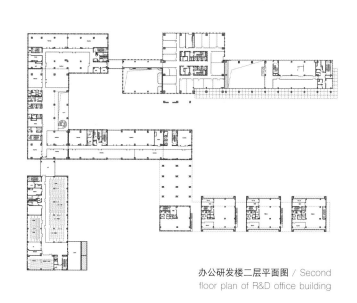

办公研发楼二层平面图 / Second floor plan of R&D office building

办公研发楼南侧立面图 / South elevation of R&D office building

04　办公研发楼，东侧入口庭院人视图 / R&D office building, view from east entrance
05　办公研发楼，东侧内广场局部透视图 / R&D office building building, perspective view from east internal square
06　办公研发楼，东侧入口庭院人视图 / R&D office building, view from the courtyard of east entrance

WUXI INTER IKEA SHOPPING CENTER

英特宜家无锡购物中心

The Wuxi Inter IKEA Shopping Center is located in the Xishan Economic Development Zone in the eastern part of Wuxi, Jiangsu Province, 8 kilometers to the east of downtown and east of the Shanghai-Nanjing Expressway. With an area of about 23 hectares and a floor area of about 23,000 square meters, the base has five floors above ground and one underground. Rising 31 meters into the sky, the project will form a large international shopping center with one-stop dining, shopping, and entertainment services.

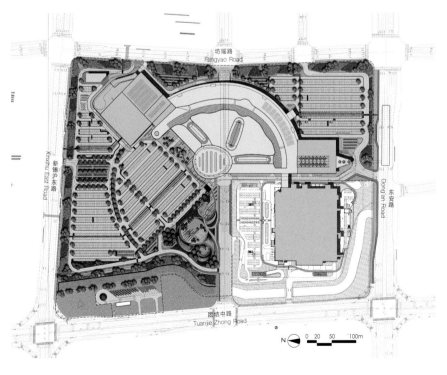

总平面图 / Site plan

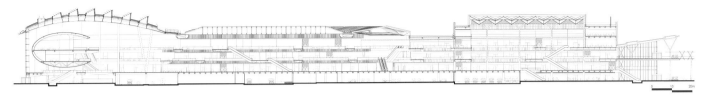

剖面图 / Section

01　中庭透视图 / Perspective of courtyard

英特宜家无锡购物中心位于江苏省无锡市东部锡山经济开发区，市中心以东8千米，沪宁高速路东侧。基地面积约23公顷，建筑面积2.3万平方米，地上5层，地下1层。地上总高度31米，项目定位于具有国际水准的区域性的大型购物中心，一站式涵盖了餐饮、购物和娱乐等功能。

该建筑的场地规划和建筑设计概念源于"蛋"的造型，不同文化中"蛋"都是繁荣和丰饶的象征。场地景观中弧形道路及建筑弧形轮廓构成醒目舒展的造型元素，在建筑造型和室内空间营造中"蛋"的形态被多次重复。

内部空间将主力店置于两端，通过中间零售餐饮业态实现连通；零售餐饮区域店铺和步行道采用曲线形式设计，优雅而充满张力；环绕中庭的步行街点缀若干蛋形休息空间，将人流引至HUB大厅，其中巨大的悬浮蛋形体POD形成瞩目的视觉中心，特殊的造型结合通高的中庭及巨大的冰场，极具戏剧性及表现力。

外立面采用简洁明快的"北欧风格"以表达来自斯堪的纳维亚半岛业主的诉求，使用大量金属装饰构件，设计细腻，色彩鲜艳，可识别性强；内外空间通透有度，虚实结合，尤其是"HUB"区，卵壳状透明的表皮，隐约映现出内部"POD"形体，空间吸引力强烈，夜晚结合照明设计呈现出的壮丽外观，使之不仅是商业内部的亮点，更是购物中心的核心。

02 总体鸟瞰图 / Overall aerial view
03 北入口透视图 / Perspective of north entrance
04 北入口大厅透视图 / Perspective of north entrance hall

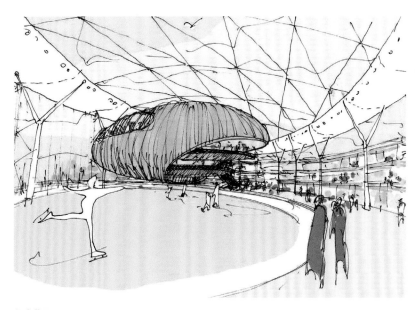

概念草图 / Sketching

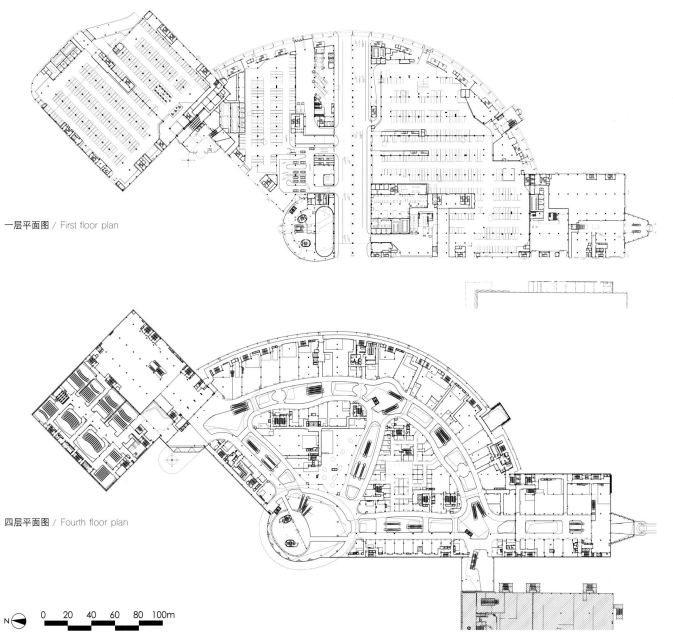

一层平面图 / First floor plan

四层平面图 / Fourth floor plan

0 20 40 60 80 100m

CRUISE TERMINAL IN THE ECOLOGICAL TOURISM ZONE OF THE YELLOW RIVER ESTUARY

黄河口生态旅游区游船码头

Located in the wetland protection area of the Dongying Yellow River Estuary in Shandong Province, the project faces the Yellow River and catches the sea at a distance. Enjoying a charming natural landscape and a superior geographical location, there are many species of animals and plants within the protected area. As a core project of the nature reserve, it will serve as a new image for Dongying City in the future, making the Dongying engineering project a landmark among Bohai Sea tourism cities.

总平面图 / Site plan

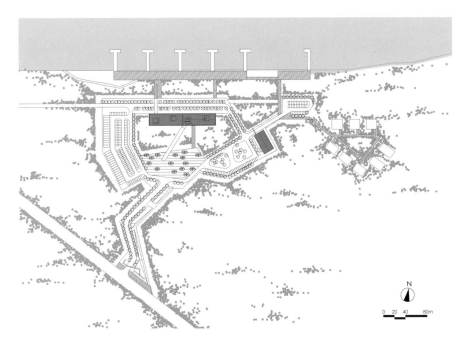

1 餐厅 / Dining
2 冷却塔 / Cooling tower
3 设备用房 / Facilities room
4 贵宾候船区 / VIP waiting room
5 候船大厅 / Waiting hall
6 综合服务台 / Integrated service desk
7 黄河文化展示区 / Display space for the Yellow River culture

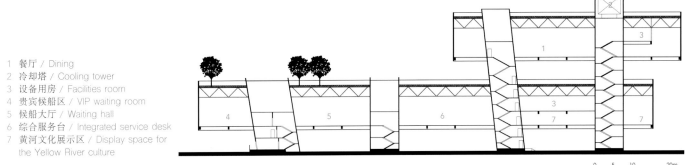

剖面图 / Section

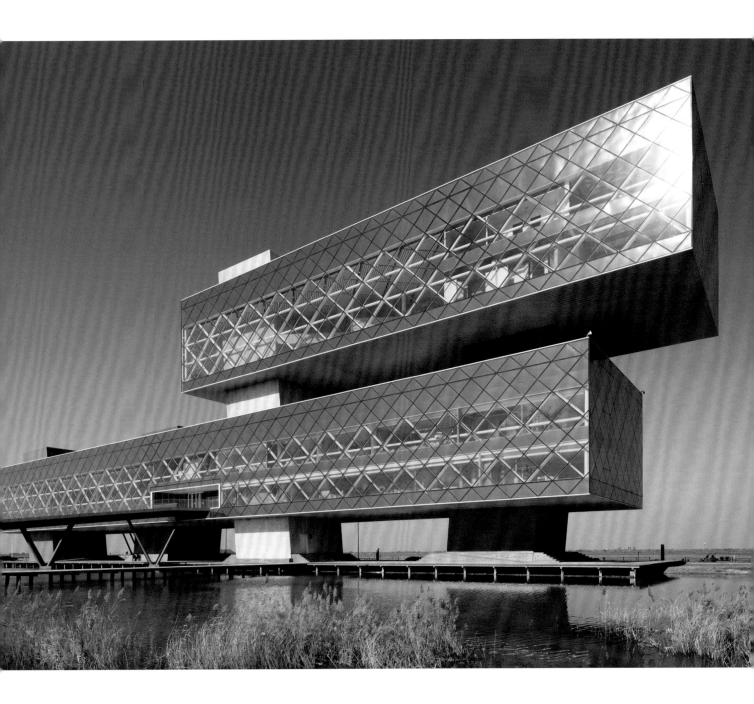

01 黄河口生态旅游区游船码头 / Cruise Terminal in the Ecological Tourism Zone of the Yellow River Estuary

概念草图 / Sketching

本项目坐落于山东省东营黄河入海口湿地保护区,面向黄河,远眺大海。自然风景优美、地理位置优越。保护区内动植物种类丰富多样。

项目占地面积约72000平方米,总建筑面积6000平方米,建筑檐口高度32米,功能子项包括:游船码头5100平方米;小木屋900平方米,建筑高度32米。项目作为自然保护区旅游开发的核心工程,建成后将成为集候船、餐饮、观光、展览于一体的综合体,并将成为东营市新的城市名片,体现东营作为沿渤海旅游城市的标志性工程。

建筑力求最小程度地破坏自然景观、整体架空、只有四个核心筒落地,整个建筑仿佛从地上长出。场地设计中除广场与停车场外,尽量保持原有湿地景观。并通过地下涵洞、廊桥的处理,使区域内外湿地环得以连通。

建筑介入自然的姿态是设计的核心话语。多层次屋顶平台为游人提供了远眺湿地景观的至高点。建筑采用创新"悬吊"结构体系,并致力于结构与建筑的整体性表达,形态现代奔放,与生态旅游区形成了强烈对比。建筑力求在这种强烈的对比中形成协调。

1 VIP候船区 / VIP waiting room
2 服务台 / Service desk
3 前室 / Prechamber
4 饮水处 / Drinking water
5 候船大厅 / Waiting hall
6 女厕所 / Lady's room
7 男厕所 / Men's room
8 储藏室 / Storage
9 入口门厅 / Entrance hall
10 感应门 / Sensor door
11 电梯厅 / Elevator hall
12 300人大餐厅 / Dining hall for 300 people
13 厨房烹饪区 / Cooking area
14 备餐区 / Pantry area

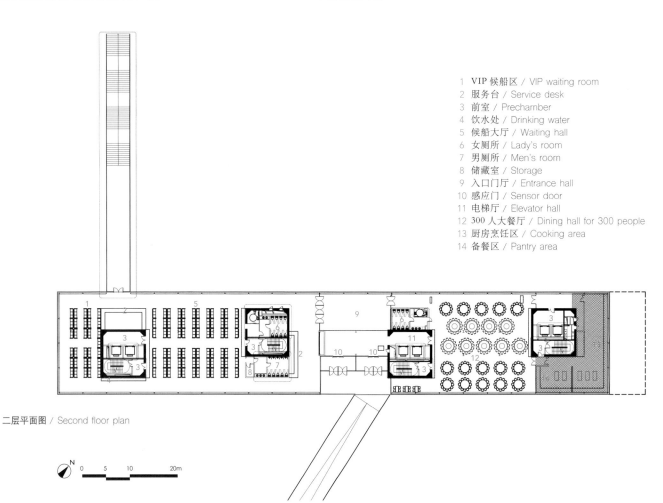

二层平面图 / Second floor plan

02 沿河透视图 / Perspective along the riverside

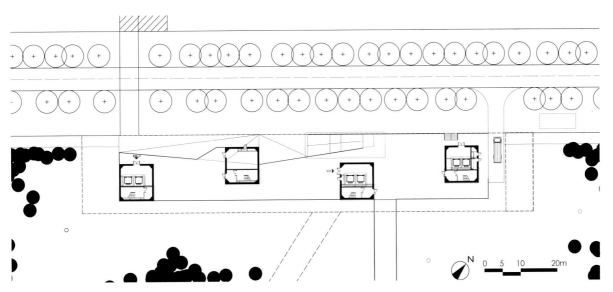

一层平面图 / First floor plan

03 夜景透视图 / Perspective of night view
04 沿河日落景观 / Sunset view along the riverside
05 室内局部 / Part of interior view

L-FERG ZHABEI PROJECT, SHANGHAI

利福上海闸北项目

The L-FERG Shanghai Zhabei Probject is the third and the largest project of the Hong Kong L-FERG International Group in Shanghai, and is located on Daning Road and Gonghe New Road, the former site of Zhabei Stadium. It neighbors Daning International to the south and Daning Park and Shanghai Circus City to the north. A shopping center, an underground high-end supermarket, an office building, and an underground garage will be built with a total floor area of about 340,000 square meters.

总平面图 / Site plan

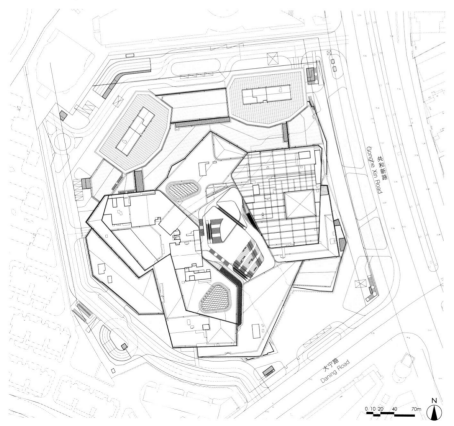

01 购物中心鸟瞰图 / Aerial view of the mall
02 购物中心室内 / Interior of the mall
03 内部庭院透视图 / Perspective of internal courtyard

利福国际集团是香港著名零售运营商，拥有全港最大型的崇光百货铜锣湾店和尖沙咀店。集团旗下现有"崇光百货""久光百货""利福广场"三个品牌。利福上海闸北项目是香港利福国际集团在上海的第三个、也是最大的项目。项目地块为原闸北体育场，位于大宁路、共和新路，南临大宁国际，北接大宁公园与上海马戏城。项目计划建造包含地上七层购物中心与久光百货、地下两层购物中心与高端超市，两幢99.9米高的5A级写字楼，地下三、四层设约1700个车位的地下车库，总建筑面积约34万平方米。

项目总体布局围绕"溪谷"的设计理念，采用"多角形"的构成手法将高层办公双塔、裙房商业及文化娱乐三大体量以环状模式进行布局。办公塔楼尽量规避了对西侧住宅的日照影响，同时又充分享用大宁公园的景观资源，两栋塔楼以对称的形式设置在面向大宁公园的基地北侧区域。商业裙房围绕基地呈"U"型布局，"U"的开口为商业的主要入口。娱乐部分及其入口设置在基地东侧靠近共和新路处。在南北高架路上即可领略到本项目大气而丰富的造型体量。同时利用基地中央大型下沉式广场，将办公、商业、娱乐有机地围合。本项目建成后将成为上海市中心北部最具吸引力的大型商业综合体。

本项目通过透水地面、屋顶绿化、节能照明、太阳能热水、雨水收集系统、节水灌溉形式、地下光导照明、二氧化碳新风监控等绿色建筑技术，以达到国家绿色二星、美国LEED金级和德国能源认证。

04　西侧立面 / West elevation
05　购物中心室内透视 / Interior perspective of the mall

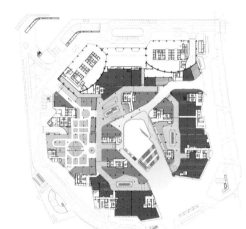
一层平面图 / First floor plan

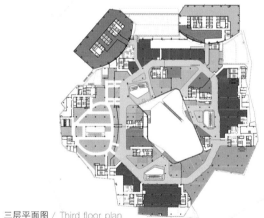
三层平面图 / Third floor plan

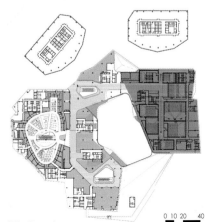
五层平面图 / Fifth floor plan

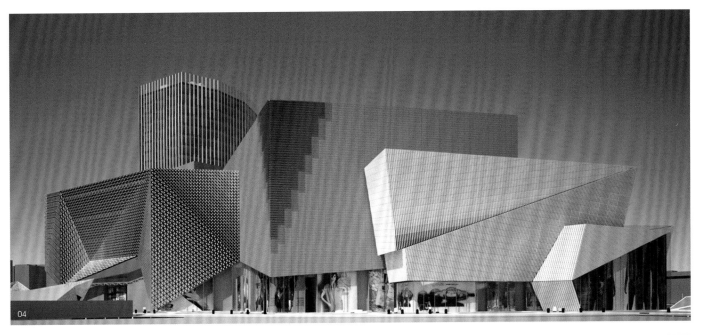

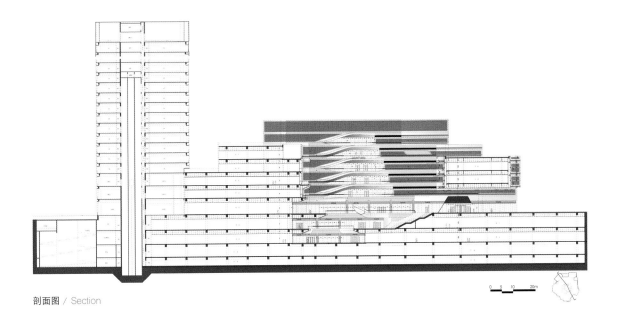

剖面图 / Section

WUHAN INTER IKEA SHOPPING CENTER
英特宜家武汉购物中心

Wuhan Inter IKEA Shopping Center, located in Etou Bay, Qiaokou District, Wuhan, faces the Third Ring Road to the west, connects to Jiefang Avenue in the south, and is adjacent to Planning Road to the north. The base has an area of about 22 hectares, a floor area of 250,000 square meters, and a total height of 39 meters above ground, with five floors.

总平面图 / Site plan

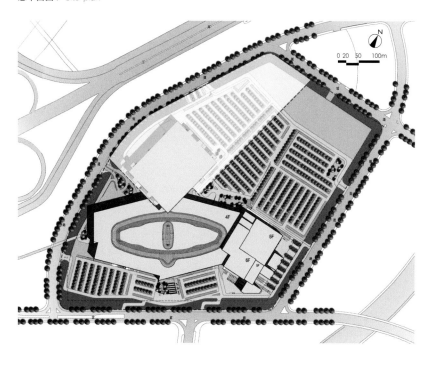

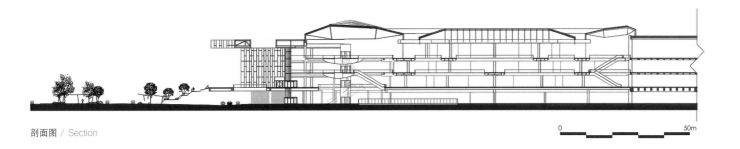

剖面图 / Section

01 购物中心与宜家家居交界处亮色体块 / Bright-colored structure at the junction of IKEA and the mall

02 购物中心入口 / Entrance to the mall
03 鸟瞰效果图 / Aerial view rendering
04 购物中心立面 / Façade of the mall

武汉英特宜家购物中心位于武汉市硚口区额头湾，西临城市三环线，南接解放大道，北近城市规划道路。基地面积约22公顷，建筑面积约25万平方米，地上5层，地上总高度39米。

宜家品牌源于和武汉同为千湖之城的瑞典，设计借鉴和吸收了武汉东湖传统园林的意向元素，设计中鱼骨状的商业平面布局和鱼鳞抽象出的建筑表皮，诠释了简洁高效的北欧商业理念。

建筑底层架空为停车场，上部为大量时尚零售、餐饮及大卖场、影院、KTV等各商业业态。建筑体量庞大，由四个入口大厅和内部商业街形成流线骨架，连接众多商业空间。室内设计通过不同的材料、几何形状及颜色来体现和强化丰富的室内感受。

外立面的斯堪的纳维亚风格以简约、功能性为特征，使用大量金属装饰构件，设计细腻，色彩鲜艳，可识别性强，成为城市新的标志。

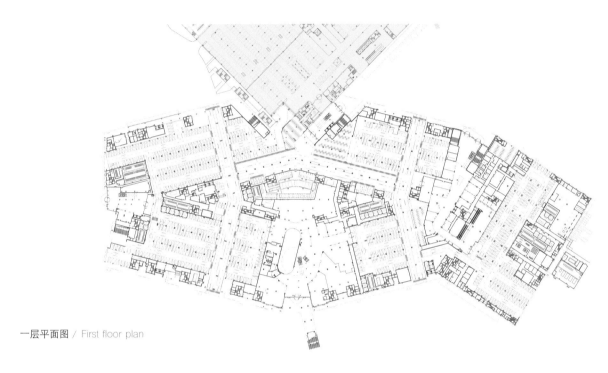

一层平面图 / First floor plan

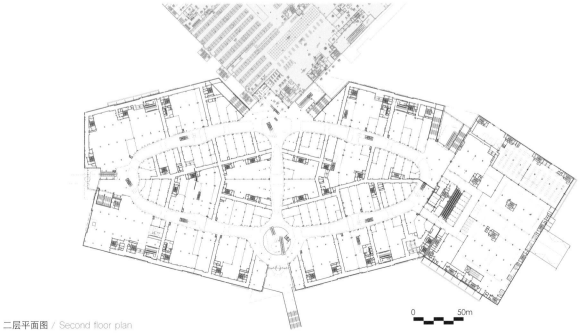

二层平面图 / Second floor plan

0　50m

WENZHOU AIRPORT TRAFFIC HUB COMPLEX

温州机场交通枢纽综合体

The Wenzhou Airport Traffic Hub Complex will become a new image of Wenzhou together with the T2 Terminal Building. An "aviation town" integrating a traffic center as well as trade, exhibition and promotion services, information exchange, innovation and promotion services, business, tourism reception, and urban complex facilities will all be constructed.

总平面图 / Site plan

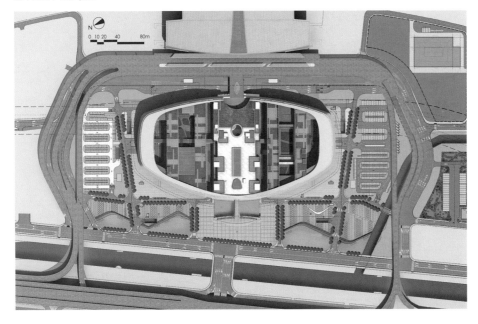

剖面图 / Section

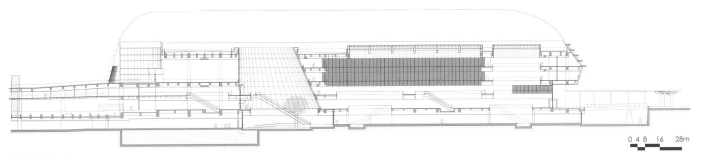

01 鸟瞰图 / Aerial view

飞翔 / Flying

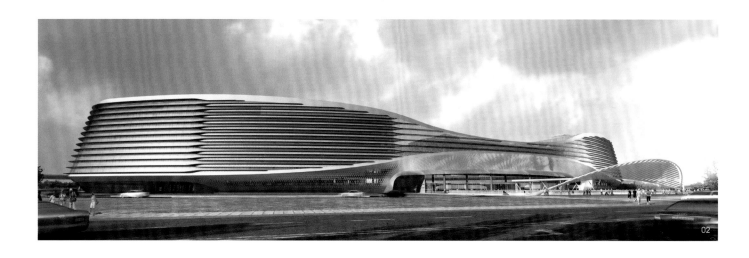

温州机场交通枢纽综合体将与T2航站楼一起，成为温州新的城市门户，合力打造一个集交通集散、商贸交易、展示推广、信息交流、创新促进、商务办公、旅游接待、城市综合配套等多种功能于一体的"航空城中城"。

温州机场交通枢纽综合体，是一次机场航站楼前区功能的突破性尝试。它打破了传统航站楼前只有停车楼单一功能的规则，对建筑功能加以发展和丰富，对航站楼原有功能加以延伸，对机场区众多服务功能进行整合，形成一个全新的、依托于机场、以机场为窗口、面向更大范围的现代服务性交通枢纽综合体。它尝试综合更多种类的交通工具，让换乘更加方便；它利用交通枢纽带来的人流，引入更丰富的服务功能，让人们的商务商贸活动更加高效；它将机场基础建设与商业开发相结合，让投资回报更加丰厚，让机场发展建设更加具有可持续发展性。

综合体位于T2航站楼前方，目前航站楼正处于施工阶段，而且出于对原有航站楼的尊重，我们采用整体式布局，与航站楼一起，采用一凹一凸的建筑造型，和谐统一，成为温州的城市新名片。

形体设计采用两端高、中部底的对称造型，中轴线处建筑高度23.7米，低于航站楼的正立面中部最高点30.5米。同时两侧高、中部底的造型，让建筑犹如展开的双翼，富有动感，也迎合了航空港交通建筑的风格特征，同时与内部功能相契合。

一层平面图 / First floor plan

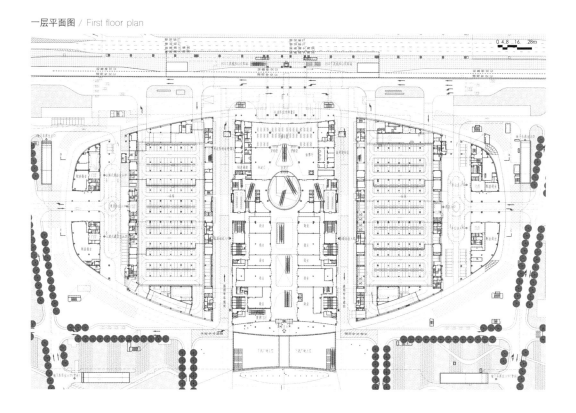

02 透视图 / Perspective
03 内院透视图 / Perspective of internal courtyard

ZHENGZHOU 27 NEW TOWER
郑州 27 新塔项目

Located in 27 New District of Zhengzhou, the tower will be a landmark complex of south Zhengzhou, and a future regional center. This includes a high-end office building, elite commercial facilities, and a five-star hotel, and the project will adopt an overall plan with two towers and commercial ancillary buildings. The tower will be built to 270 meters.

总平面图 / Site plan

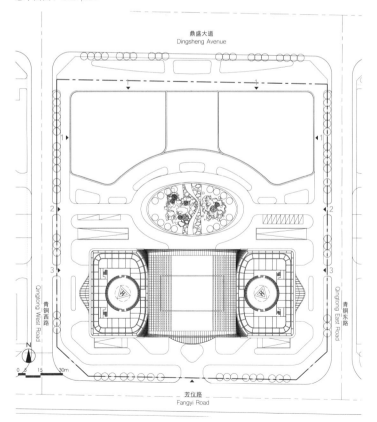

1　餐饮入口 / Entrance to dining area
2　车行出入口 / Entrance for vehicles
3　办公出入口 / Entrance for staff

01 鸟瞰图 / Aerial view

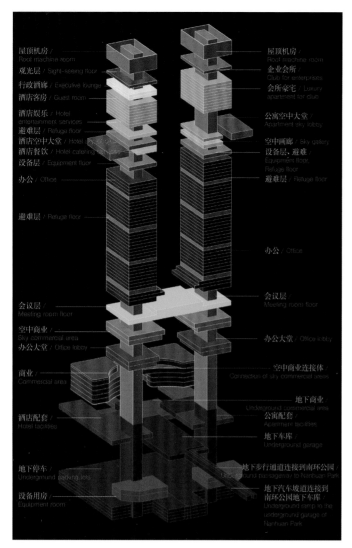

功能分析图 / Function analysis diagram

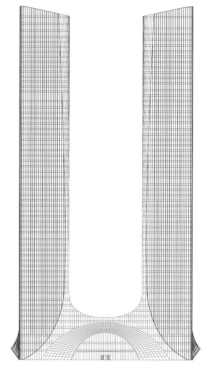

南侧立面图 / South elevation

项目位于郑州市 27 新区。建成后将作为郑州南部地标性城市综合体，未来的区域中心。项目包含甲级办公、精品商业和五星级酒店。方案采用双子塔模式辅以商业裙房的总体规划，塔楼总高度 270 米。

超高层建筑具有极强的类型特征，是否能够发现超越固有模式的崭新突破？针对问题的思考和探寻贯穿于该类型建筑的设计实践，郑州 27 新塔项目的设计策略作为问题的回应，提出了一种新颖的超高层双子塔模式。

建筑尝试全新的双塔体量原型，提出拱形结构衔接双塔的形体策略，打造崭新独特的建筑形象。塔楼造型控制遵循刚柔并济的设计理念，以方形体量为基础，通过削切、柔化等形体处理手段，化方为圆，进而与底部拱形结构融为一体。整个设计过程结合理念创新、功能要求以及建筑技艺的探索。

双塔的设计意象——"中原之虹"，体现了建筑独有的造型特征，也表达了设计者在作品中所蕴藏的美好期许。

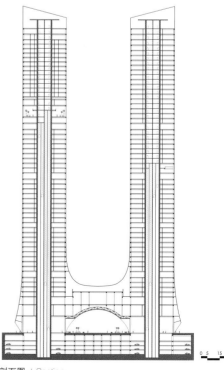

剖面图 / Section

02 人视图 / Perspective view

1 商业后勤通道 / Passage way for support staff
2 酒店办公人员出入口 / Entrance for hotel staff
3 商业后勤内院 / Cortile for support staff
4 场地西侧车行出入口 / Entrance for vehicles to the west of the site
5 大巴车停车位 / Parking lots for buses
6 小车停车位 / Parking lots for cars
7 出入口（办公、商业）/ Entrance (office & commercial)
8 场地东侧车行出入口 / Entrance for vehicles to the east of the site
9 酒店出入口 / Entrance to hotel
10 宴会厅入口 / Entrance to banquet hall
11 大巴车下车落客 / Drop-off area for buses
12 出租车等候区 / Waiting area for taxis
13 地下车库出入口（酒店）/ Entrance to underground garage (hotel)
14 地下车库出入口（办公）/ Entrance to underground garage (office)
15 环岛景观 / Landscape island
16 场地南侧车行出入口 / Entrance for vehicles to the south of the site
17 办公出入口 / Entrance to office

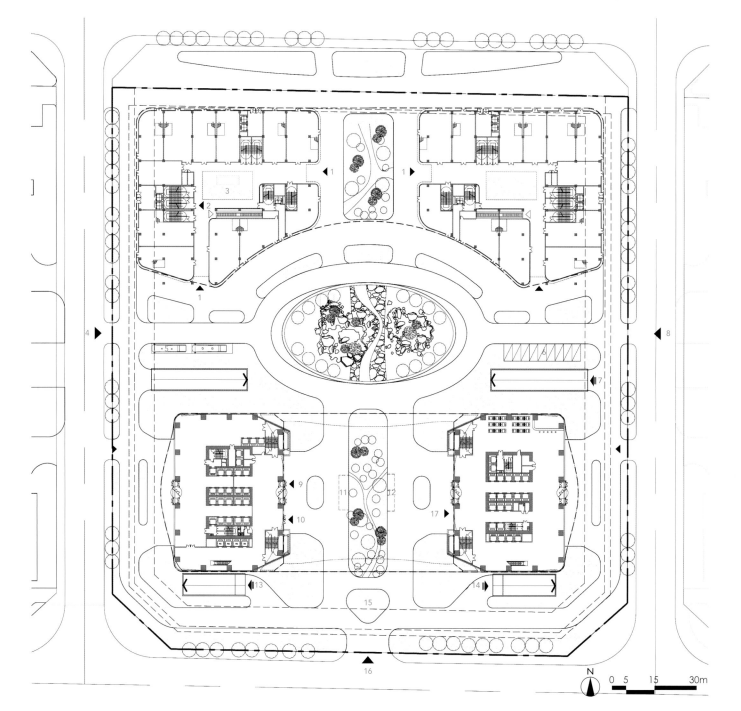

一层平面图 / First floor plan

03 人视图 / Perspective view

东塔9-17层平面图 / 9th to 17th floor plan of East Tower

1 空调机房 / Air conditioning facilities room
2 候梯厅 / Waiting hall
3 前室 / Prechamber
4 强电间 / High voltage room
5 电缆间 / Cable room
6 弱电间 / Low voltage room
7 茶水间 / Tea room
8 水管间 / Pipe room

PLOT NO. 4 OF KUNMING DIANCHI INTERNATIONAL CONVENTION AND EXHIBITION CENTER (MAIN TOWER)
昆明滇池国际会展中心 4 号地块（主塔）

Plot No. 4 of Kunming Dianchi International Convention and Exhibition Center is located to the south of downtown, and northwest of Chenggong New District, adjacent to the main building and the regional axis of the convention and exhibition center. To the south of the project are the broad waters of Dianchi Lake, with West Mountain to the far west, and the Wetland Park in the east, with the building enjoying a rich landscape. The project includes high-end office buildings and hotels.

总平面图 / Site plan

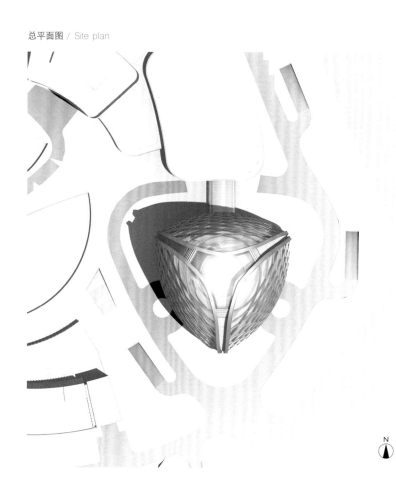

01 西北人视图 / View from northwest

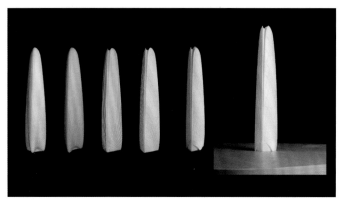

设计过程的模型演变 / Model evolution to demonstrate the design process

昆明滇池国际会展中心 4 号地块位于主城区以南、呈贡新区西北，紧临会展中心主体建筑和区域中轴线。项目南向具有滇池一线开阔水景，西向可远眺西山山景，东临湿地公园，景观资源丰富。建筑功能包括高端写字楼和酒店。

"孔雀之翼"：设计灵感来源于云南的地域标志——孔雀，将其巧妙地融合于超高层立面中，如滇池水畔的孔雀开屏，成为昆明的城市新地标。

"临水之峰"：在滇池水边设立单栋最高的标志性主塔，统领一组形态多姿的建筑群，柔中带刚的轮廓是水和山的性格表现。高低起伏、主次分明的天际线宛如临水的山峰，演绎滇池周边连绵起伏的自然景观。

超高层主塔矗立在地块东南角，造型圆润饱满，拥有 360 度景观面；形体收分有致、刚柔相济，既符合超高层塔楼如山般挺拔的形象要求，又与滇池水的柔美相得益彰；立面线条呼应塔楼形体，交错而上，不仅巧妙地结合了结构元素，也使"孔雀"的概念得以体现。

结构体系采用带加强层的钢筋混凝土框架核心筒体系。楼面体系由钢框架梁、钢次梁以及组合楼板构成，设置粘滞阻尼伸臂桁架 + 环带桁架，受力合理，造价经济，减震效果好。曲面幕墙通过参数化计算，94% 的玻璃板块优化为平板玻璃。局部双曲面板块还可通过特殊构造、局部镂空等做法简化为单曲面板块，节省幕墙造价，降低施工难度。设计应用 BIM 技术，全面实现项目信息的无障碍交流。

02 整体鸟瞰图 / Overall aerial view

一层平面图 / First floor plan

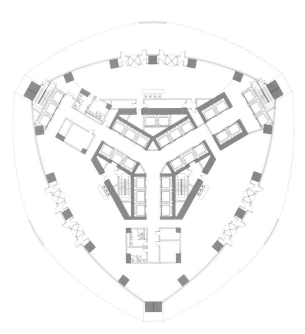

三层平面图 / Third floor plan

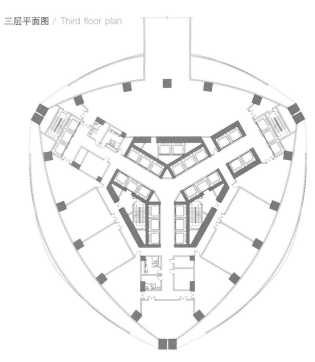

酒店空中大堂层平面图 / Sky lobby floor plan

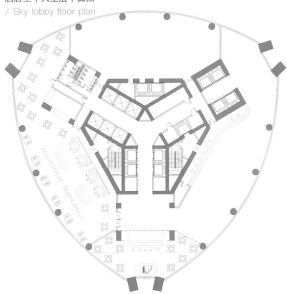

酒店餐厅层平面图 / Dining hall floor plan

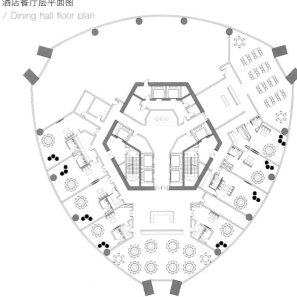

酒店低区客房层平面图 / Guest room floor plan for lower floors

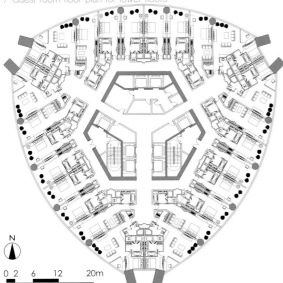

办公低区平面图 / Office floor plan for lower floors

N
0 2 6 12 20m

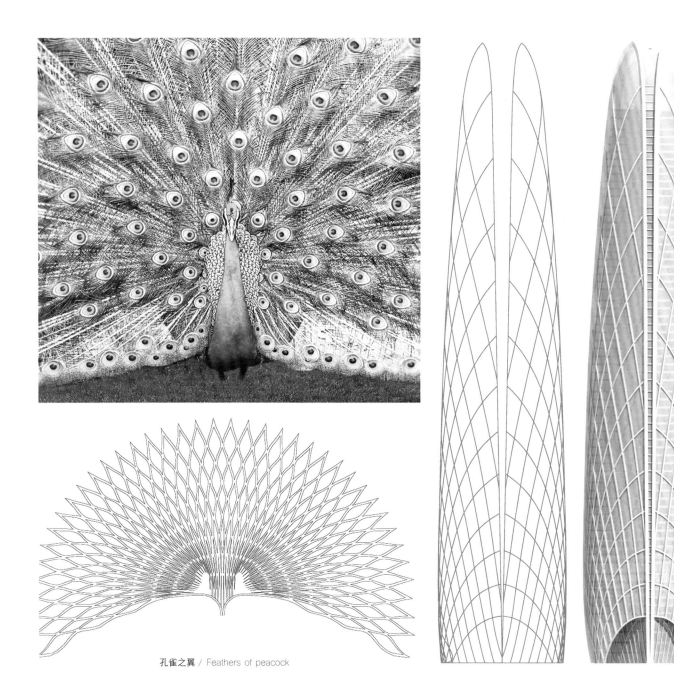

孔雀之翼 / Feathers of peacock

北侧立面图 / North elevation

南侧立面图 / South elevation

剖面图 / Section

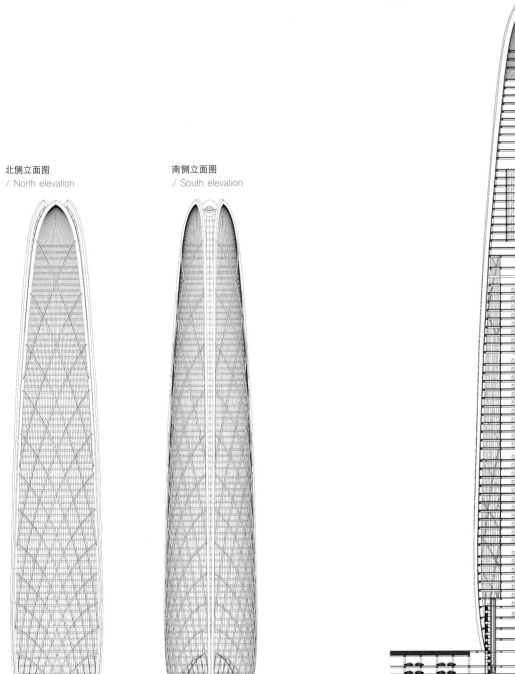

SUINING SPORTS CENTER

遂宁市体育中心

Located in Hedong District of Suining City, Sichuan Province, Suining Sports Center meets the full standards for comprehensive provincial sports competitions, national sporting events, and public fitness venues. Built on an area of 12.75 hectares with a gross floor area of 79,741 square meters, the sports center has a stadium, swimming pool, and outdoor warm-up areas, as well as training and activity venues.

01 夜景局部透视图 / Perspective of night view

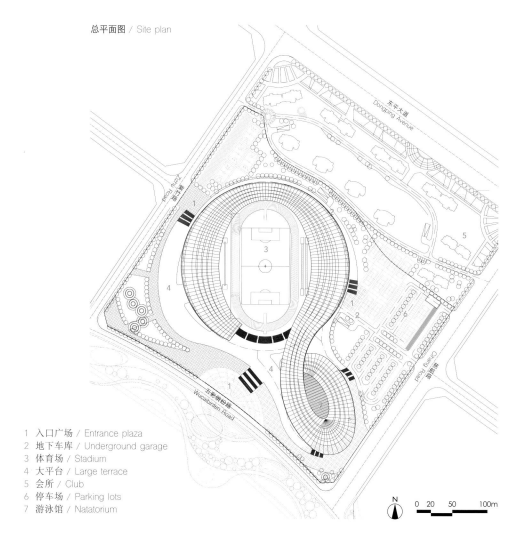

总平面图 / Site plan

1 入口广场 / Entrance plaza
2 地下车库 / Underground garage
3 体育场 / Stadium
4 大平台 / Large terrace
5 会所 / Club
6 停车场 / Parking lots
7 游泳馆 / Natatorium

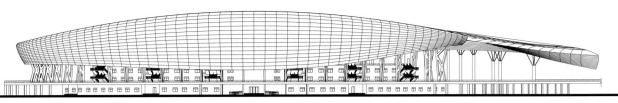

西侧立面图 / West elevation

遂宁市体育中心位于四川省遂宁市河东区，功能为满足该地区承办省级综合性运动会、全国性单项比赛及全民健身等需求。基地面积12.75公顷，总建筑面积79741平方米，包含体育场、游泳馆以及配套的室外热身、训练及活动场地。

攫取于当地的文化背景，结合与建筑的形式语言，遂宁体育中心从城市文化内核衍生出舞动、双环等设计概念。从内使建筑植根于城市大背景下，成为"土生土长"的城市一员。

回归体育建筑的本质亦即处理结构与形式的二元关系，两者即互相依存又矛盾对立。结构既是骨架又是外在张力，形式既可以是外表也可以是内在的力学原理。发现形式与结构的平衡点和结合点一直是体育建筑设计的重点。

遂宁体育中心的形式源自于当地文化元素的提取，体育场和游泳馆统一在同一片屋顶下。屋顶起自于体育场的一隅，顺势以圆形环绕场地一周，后停顿于游泳馆，再以一捺之势包裹游泳馆，最终收于两形之停顿处。整个曲线屋顶流畅自然地表达了舞动的概念。顺应屋顶的造势，两馆之间又以曲线形大平台相连接，完成了整体化的形式表达。与此相配合的，遂宁体育中心选择了顺应屋顶形式的钢桁架结构。

为了适应场馆自身可持续运营的需求，设计师在设计过程中加入了可综合利用的因素。体育中心平台下的空间中穿插式布置了休闲体育区、健身俱乐部、体育纪念品和器材商店等大众体育设施，同时也加入了茶楼、酒吧、咖啡吧、商业、按摩等提高盈利可能性的商业性设施，在最大程度上给场馆的后续运营带去多种可能性，从而使建筑空间能物尽其用，统一赛时和赛后的二元关系，长久地可持续发展下去。

02 日景鸟瞰图 / Aerial view

北侧立面图 / North elevation

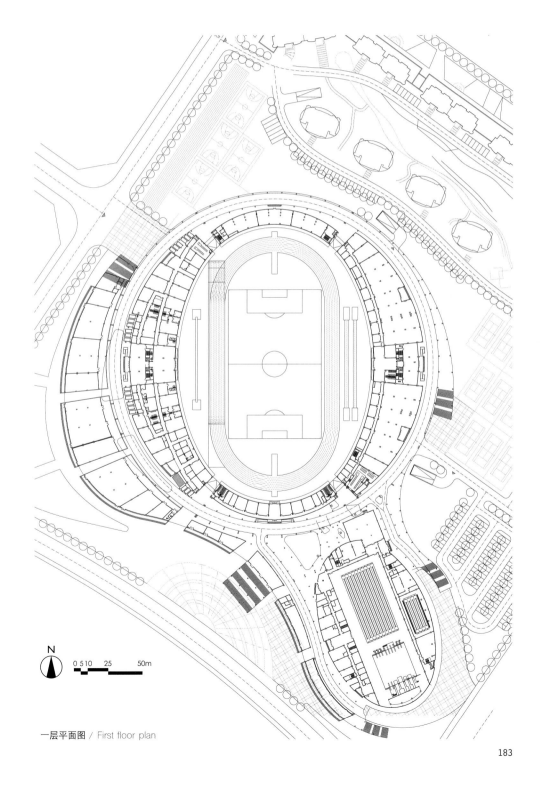

一层平面图 / First floor plan

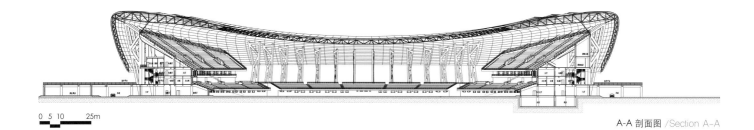

A-A 剖面图 /Section A-A

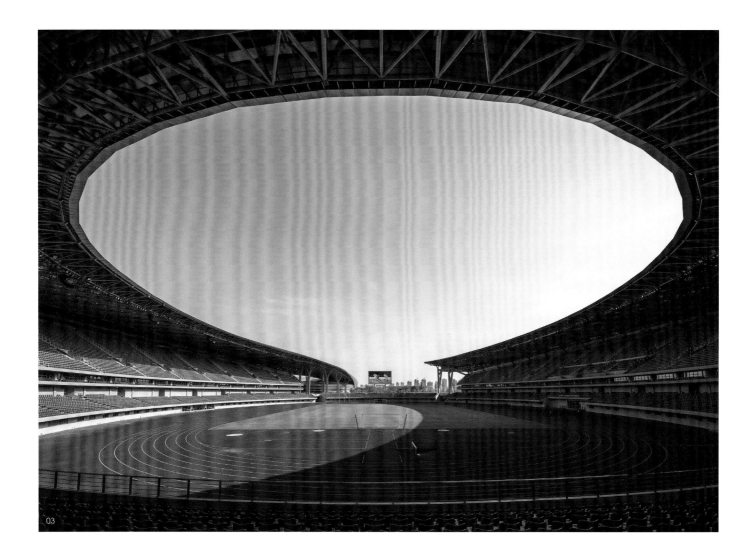

03 体育场内场 / Stadium infield
04 游泳馆泳池区 / Swimming pool of the natatorium
05 体育场室内 / Stadium interior
06 游泳池跳板 / Springboard of the swimming pool

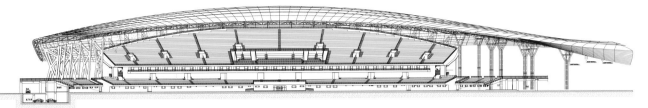

B-B 剖面图 / Section B-B

185

JINING OLYMPIC SPORTS CENTER

济宁奥体中心

The Jining Olympics Sports Center was the main venue of the Shandong Provincial Sports Meeting in 2014, and offered major support to the event with its one field and four halls. The sports center possesses a fully functional stadium, swimming pool, shooting hall, and outdoor sports field, presenting a radial layout.

总平面图 / Site plan

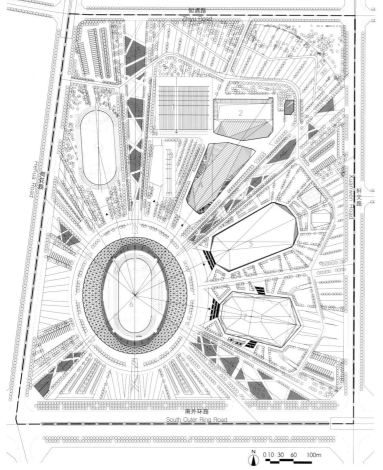

1 室外射箭场（预赛）/ Outdoor archery range (for preliminary contest)
2 室外射击场 / Outdoor shooting range
3 室外射箭场（决赛）/ Outdoor archery range (for finals)
4 射击射箭馆 / Shooting and archery pavilion
5 游泳跳水馆 / Swimming and diving pavilion
6 文体中心（已建）/ Cultural and sports center (built)
7 综合体育馆 / Complex gymnasium

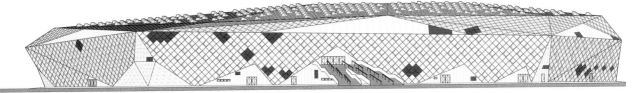

综合体育馆北侧立面图 / North elevation of complex gymnasium

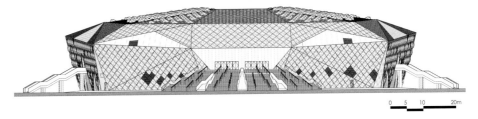

综合体育馆西侧立面图 / West elevation of complex gymnasium

01 总体鸟瞰图 / Overall aerial view

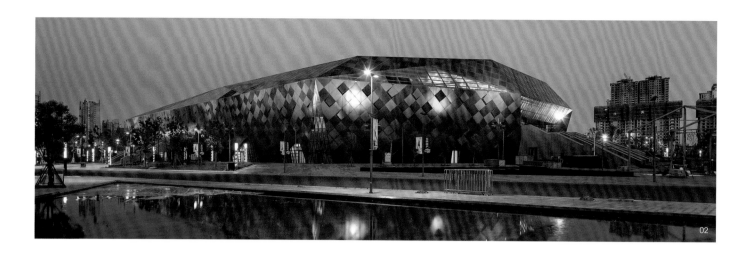

济宁奥体中心项目是2014年山东省省运动会举办主场地，内部一场四馆将为赛事提供主要支撑，在总体上综合体育馆、游泳馆、射击馆、室外运动场等，围绕体育中心呈放射式布局。

该项目功能完备，拥有一流的体育使用功能、一流的体育竞赛设施、一流的体育竞技环境，可以满足国际国内体育比赛的要求。

奥体中心建筑体除了综合效应，在设计上超前谋划、周密安排，建筑空间的有效利用，可适应赛后多种功能的使用要求和可经营性。在使用上，既要建设成为举办国内外大型比赛的竞技中心，满足大型运动会及国际单项比赛的要求，又要建设成为济宁市广大市民的健身、休闲、娱乐的载体。

建筑形象具有时代特征，代表了独特的地域文化特色，并以现代的设计理念展示具有现代风貌的体育建筑形象和城市地域文化特色。体育中心的建筑形象，不仅作为生态建筑，也作为新济宁的标志为城市建设增添新的气息，将成为济宁市独具个性的标志性建筑。

建筑设计整体和谐。项目实施的指导原则为"经济实用、适度超前、功能齐全、朴实大方，国内一流"，能满足"节地、节水、节能、节材，促进循环经济发展"的使用要求，满足规定的各种规范、指标要求，采用成熟的技术和已经验证过的设备，满足国内规范约束下的实现可能性。项目生命周期内便于经营与维修，使项目最终实现"功能完善、环境优化、特色鲜明、以人为本、模式创新、过程经济、项目最优"的建设目标。

综合体育馆一层平面图 / First floor plan of complex gymnasium

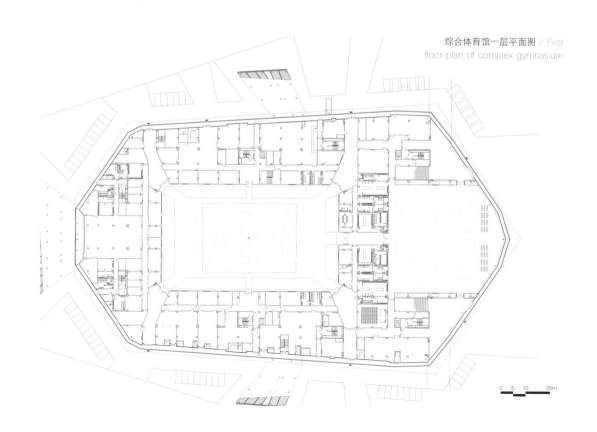

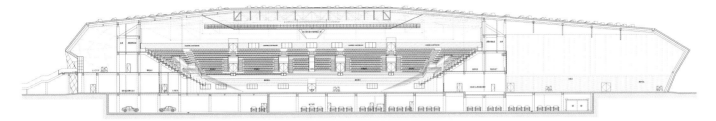

综合体育馆剖面图 / Section of complex gymnasium

02　综合体育馆夜景 / Night view of complex gymnasium
03　综合体育馆室内 / Interior of complex gymnasium

射击射箭馆立面图 / Elevation of shooting and archery pavilion

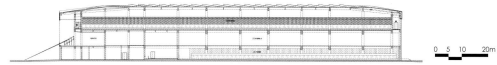

射击射箭馆剖面图 / Section of shooting and archery pavilion

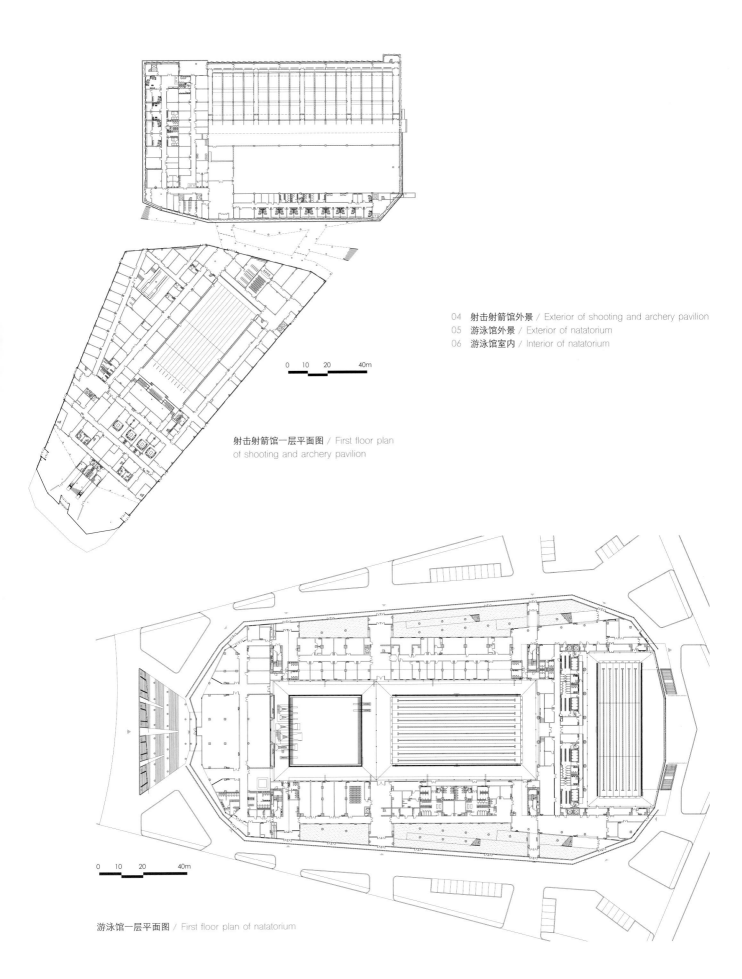

射击射箭馆一层平面图 / First floor plan of shooting and archery pavilion

04 射击射箭馆外景 / Exterior of shooting and archery pavilion
05 游泳馆外景 / Exterior of natatorium
06 游泳馆室内 / Interior of natatorium

游泳馆一层平面图 / First floor plan of natatorium

GYMNASIUM OF CHANGSHU SPORTS CENTER

常熟市体育中心体育馆

Located east of the axis of the cultural area of Changshu, the gymnasium and swimming pool of the Changshu Sports Center were built and put into operation in 2003. With an area of 30,436 square meters, the gymnasium offers a venue for handball, indoor soccer, basketball, volleyball, and other competitions and training events. Meanwhile, functions beyond competition were taken into consideration during the design to achieve maximum levels of sustainability.

总平面图 / Site plan

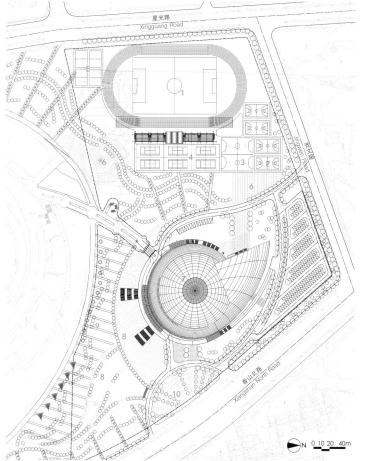

1	训练场 / Training ground	6	休闲广场 / Leisure plaza
2	篮球场 / Basketball court	7	体育馆 / Gymnasium
3	5人制足球场 / Five-a-side football field	8	运动广场 / Sports plaza
4	网球场 / Tennis court	9	亲水平台 / Water terrace
5	门球场 / Gateball court	10	入口广场 / Entrance plaza

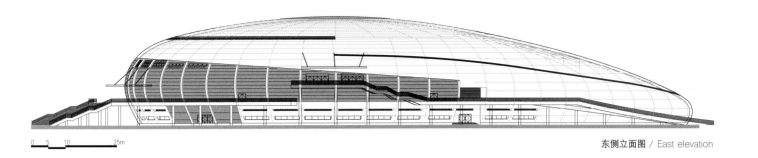

东侧立面图 / East elevation

01　日景透视图 / Perspective view

常熟市体育中心位于常熟市文化片区中轴线东部，其中体育场与游泳馆已于2003年建成并投入使用。体育馆总建筑面积为30436平方米，可进行手球、室内足球、篮球、排球等比赛和训练。同时也对赛后利用进行了考虑，最大限度的将功能复合地整合入设计之中，实现可持续利用。

整体性、一体性、复合性，从大到小、由表及里，很好地诠释了常熟市体育馆的设计思考。整体性实现了场地内三位一体的平衡关系，使体育馆植根于场地环境内。一体性则探求了体育建筑本体的设计逻辑——结构与形式的平衡关系，达成了内外交融的形式结构表达。更为重要的是，常熟市体育馆既满足当下，又兼顾未来，用复合性满足体育馆未来的可持续发展。

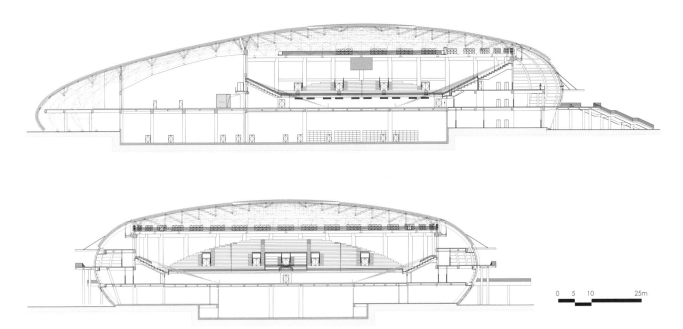

剖面图 / Section

02 东侧外景 / Exterior of east elevation
03 整体透视图 / Overall perspective
04 外景局部 / Part of exterior view

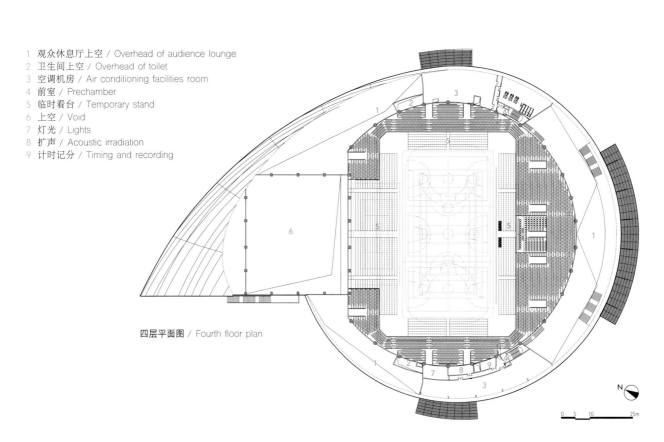

1 观众休息厅上空 / Overhead of audience lounge
2 卫生间上空 / Overhead of toilet
3 空调机房 / Air conditioning facilities room
4 前室 / Prechamber
5 临时看台 / Temporary stand
6 上空 / Void
7 灯光 / Lights
8 扩声 / Acoustic irradiation
9 计时记分 / Timing and recording

四层平面图 / Fourth floor plan

05 入口局部 / Part of the entrance
06 室外局部 / Part of the outdoor space
07 室内屋顶 / Interior rooftop

SHANGHAI CHONGMING SPORTS TRAINING BASE
上海崇明体育训练基地

Located in Chenjia Town, Chongming District, the sports training base stretches to 55 Tanghe in the northeast, Beiyan Road in the south, and to a reserved development zone in the west. Phase I has taken up 558,920.6 square meters with a gross floor area of 189,708 square meters.

总平面图 / Site plan

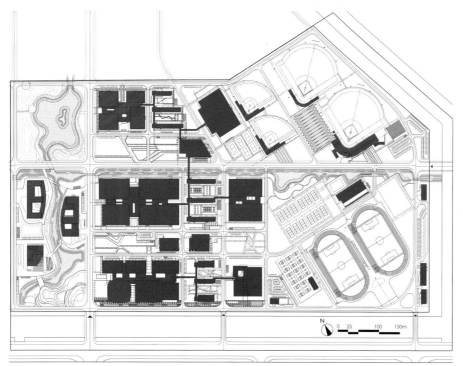

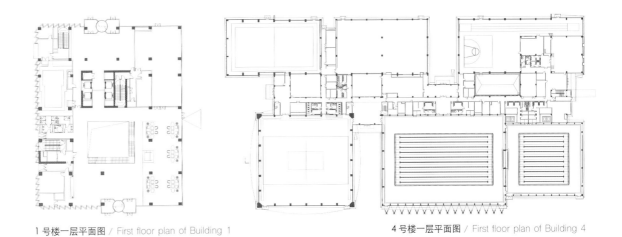

1号楼一层平面图 / First floor plan of Building 1　　　　4号楼一层平面图 / First floor plan of Building 4

01　总体鸟瞰图 / Overall aerial view

上海崇明体育训练基地项目选址为上海崇明地块，位于崇明区陈家镇，东北至55塘河、南至北沿公路、西至规划发展预留用地。一期建设用地面积558920.6平方米，一期总建筑面积为189708平方米。

上海崇明体育训练基地的定位是：建成符合具有国际竞争力和影响力的国际体育强市——上海市体育事业发展目标；突出奥运战略，满足并提升上海优秀运动团队日常训练需求，同时为部分项目国家队备战奥运提供保障，满足高水平运动团队交流训练比赛的需求，满足部分运动项目规则对场馆有特殊设置的竞赛需求；构建"训练、科研、医疗、教育"为一体，国际先进、国内一流、高科技、多功能的现代化国家级体育训练基地。

项目整体设计注重营造适应环境与环境互动的建筑空间，以保证训练基地中良好的空间环境，避免过大体量带来的压迫感和宏观室外空间尺度带来的紧迫感，营造舒适宜居的生态型训练基地环境。

项目总体布局主要分为中轴公共建筑区、第一训练组团、第二训练组团、第三训练组团以及后勤辅助用房。其中中轴公共建筑区包含有运动员公寓及管理楼、科研医疗楼和教学楼，可满足基地办公、接待、医疗、教学等需求；第一训练组团坐落于基地东端南侧，由游泳、水疗、综合教学比赛、身训和篮球馆；第二训练组团坐落于东端中部，由艺术体操、蹦床、体操训练和网球馆组成；第三训练组团坐落于东端北部，由跆拳道、拳击、武术、击剑、现代五项馆和手球馆组成。各训练组团均分别设置了运动员公寓。基地东侧室外训练场地中根据训练需求设置现代五项跑射馆、棒球比赛场辅助用房和室外场地附属用房。在各训练组团之间设置有两个食堂，主要供应运动员、教练员、教育科研人员和后勤人员的餐饮需求。

1号楼南立面图 / South elevation of Building 1

1号楼剖面图 / Section of Building 1

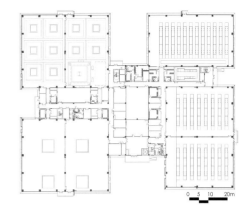

6号楼一层平面图 / First floor plan of Building 6

02 公共服务设施鸟瞰图 / Aerial view of public service facilities
03 1号楼透视图 / Perspective of Building 1

04 4号楼透视图 / Perspective of Building 4
05 4号楼游泳馆室内效果图 / Natatorium interior rendering of Building 4
06 6号楼透视图 / Perspective of Building 6

4号楼南侧立面 / South elevation of Building 4

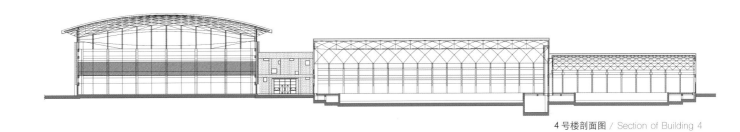

4号楼剖面图 / Section of Building 4

6号楼南侧立面 / South elevation of Building 6

RECONSTRUCTION OF NINGBO RAILWAY STATION

改建铁路宁波站改造工程

Ningbo Railway Station is located in Haishu District, the downtown area of Ningbo, Zhejiang Province, and is enclosed by Nanzhan West Road, Cangsong Road, Yongshuiqiao Road, and Sanzhi Street. It is the "core" traffic hub of the "one core, one shaft, and six clusters" of the planned urban spatial structure of Haishu District.

总平面图 / Site plan

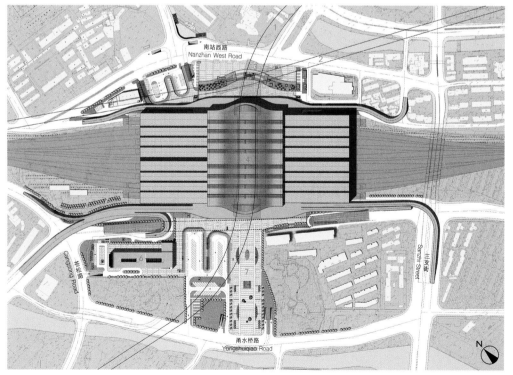

1. 地铁2号线 / Subway Line 2
2. 地铁4号线 / Subway Line 4
3. 北广场 / North plaza
4. 国铁站房 / National Railway Station building
5. 永达路下立交 / Under crossing of Yongda Road
6. 市域公交车站 / City bus stop
7. 南广场 / South Plaza

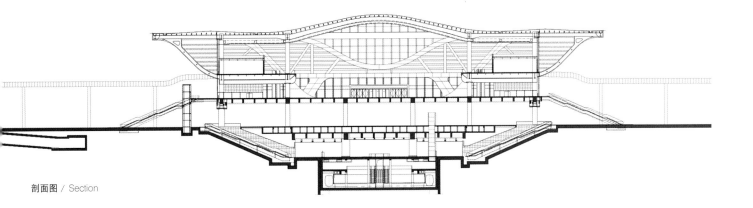

剖面图 / Section

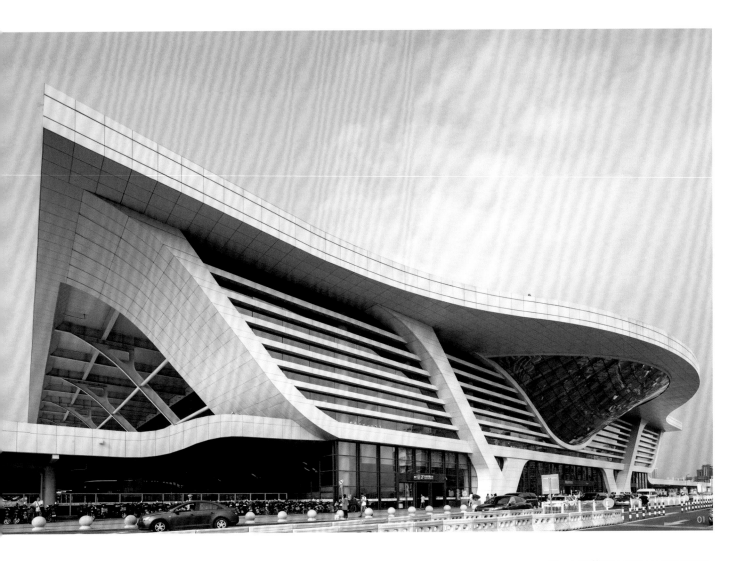

01 入口透视图 / Entrance perspective

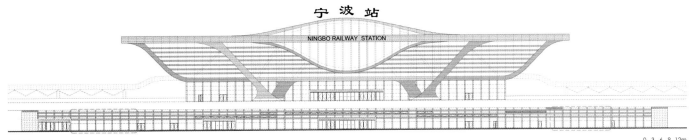

南侧立面图 / South elevation

铁路宁波站交通枢纽位于浙江省宁波市中心城区——海曙区,枢纽区由南站西路、苍松路、甬水桥路、三支街围合而成,是海曙区规划的"一核、一轴、六组团"城市空间结构中的"一核",即交通中心区。建筑用地面积22.18万平方米,总建筑面积11.96万平方米。地上两层,地下一层。地下室下方为宁波市轨道交通地铁2号线、4号线,两者一同设计,同时施工建造,地上总高度39.1米,总埋深25.5米。

宁波站的设计理念为"天一生水",出自中国易经,取意"天一生水,水生万物"。

宁波站的建筑构思由位于建筑中央的一滴晶莹剔透的"水滴"幻化而成,"水滴"既使建筑的进站广厅宽敞明亮,又使建筑的外部造型成为城市景观的汇聚点,无论白天夜晚都有强烈的中心感。同时,随着"水滴"的扩散,形成优美的波浪造型,构成建筑屋顶的造型意向。

"天一生水"既展现了宁波的地域性文化,也具有鲜明的时代感。正是得益于"天一生水,水生万物"哲学思想的启发,使宁波走上了"海上丝绸之路"的巅峰。

宁波站站房地处城市中心区,作为城市大型综合交通枢纽重要节点,高度整合了市政交通设施,相接的两条不同方向的城市轨道交通线,一纵一横分别从南北、东西方向穿越站房的腹地和北侧广场,结合北侧广场下方的甬达路下穿地道,东西向贯穿铁路客站,枢纽区地下空间城市交通立体接壤,充分满足城市区域交通疏解的长远发展需求。

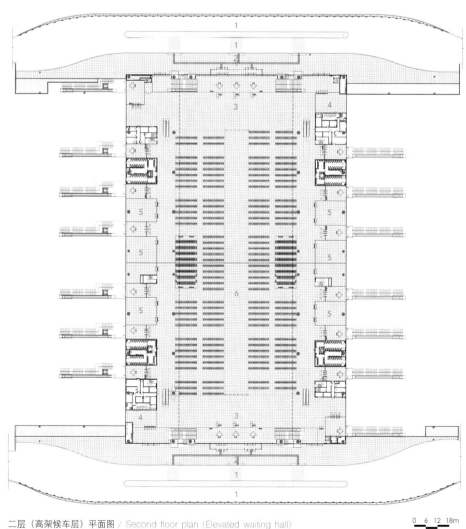

1 行车带 / Pass way for cars
2 落客平台 / Dropping-off terrace
3 进站广厅 / Entrance hall
4 售票厅 / Ticket hall
5 旅客服务 / Passenger service
6 高架候车大厅 / Elevated waiting hall

二层(高架候车层)平面图 / Second floor plan (Elevated waiting hall)

02 东南侧鸟瞰图 / Aerial view from southeast
03 候车大厅透视图 / Perspective of waiting hall
04 入口室内透视图 / Perspective of entrance interior

STATION BUILDING OF DALIAN NORTH STATION OF THE HARBIN-DALIAN PASSENGER LINE

哈大客专大连北站站房工程

Formerly known as New Dalian Station, Dalian North Station is a sub-project of the Harbin-Dalian Passenger Line, which is located in Ganjingzi District, Dalian City, with 10 platforms and 20 lines. The station building has an area of 68,965 square meters, and will focus on the passenger transportation capabilities of the railway to integrate city rail transit, urban public transport, taxies, and private vehicles, as well as all sorts of other traffic facilities and transport modes.

总平面图 / Site plan

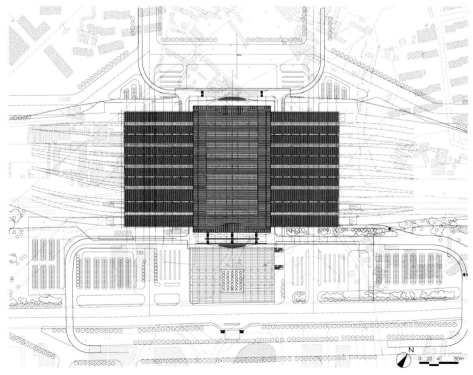

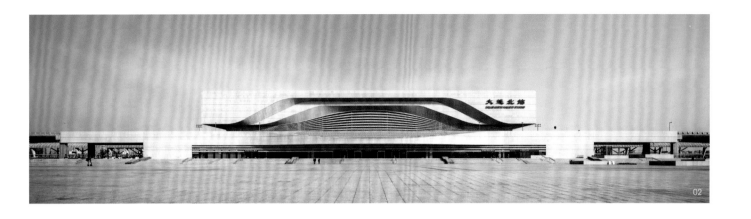

01 北广场夜景 / Night view of north plaza
02 北广场主立面图 / Façade on the side of north plaza

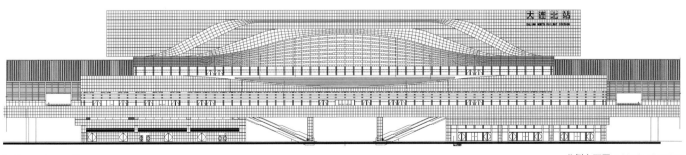

北侧立面图 / North elevation

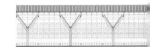

大连北站原称新大连站,为哈大客专工程建设项目的分项。基地位于大连市甘井子区,车场规模为10个台面20线,站房面积68965平方米。大连北站将建设成为以铁路客运为中心,集城市轨道交通、市区公交、出租车以及社会车辆等各种交通设施及交通方式的客运综合交通枢纽。

大连北站在总体设计上采用"一心、两轴、四区"的总体规划格局;站区交通疏解采用"两纵两横一通道"的模式;规划的城市轨道交通1、2、4号线在大连北站北侧平行通过。出站人流交通组织集中设于站台层下方24米宽的城市南北通廊步行区,并方便抵达车站南、北广场,接驳地铁、公交、出租车以及社会车场。

大连是我国最重要的北方城市之一,同时也是环渤海湾最著名的港口城市和度假胜地,因此,大连城市也以其"雄浑大气"和"浪漫诗意"的双重特征而著称。为充分展现大连的城市精神和时代风貌,建筑以一块被"海水雕琢的巨石"作为形态构成创意源,寓意大连城市"刚柔相济"的双重特征。

根据主题立意,大连北站的建筑形态一方面以"舒展、方正"的巨石结构造型,另一方面,东西向延展的站台雨棚,采用"波浪"造型,随着海水的波动,巨石的中央被海水雕琢出来的弧形空间,优美、俊秀,一气呵成,表现出大连城市"浪漫诗意"的文化气质。

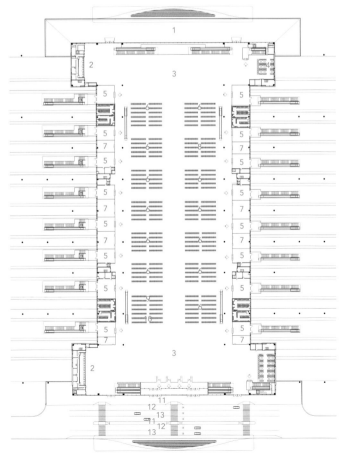

03　雨棚透视图 / Canopy perspective
04　高架候车厅 / Elevated waiting hall
05　出站大厅 / Exit hall
06　检票进站口 / Ticket entrance

高架层平面图 / Floor plan of elevated waiting hall

1　屋顶平台 / Roof terrace
2　售票厅 / Ticket hall
3　进站广厅 / Entrance hall
4　软席候车区 / Soft seats waiting area
5　进站通道 / Entrance pass way
6　基本候车区 / Waiting area
7　旅客服务 / Passenger service
8　母婴候车区 / Waiting area for maternal and infants
9　军人候车区 / Waiting area for soldiers
10　团体候车区 / Waiting area for groups
11　落客平台 / Dropping-off terrace
12　停车带 / Lay-by
13　行车带 / Pass way for cars

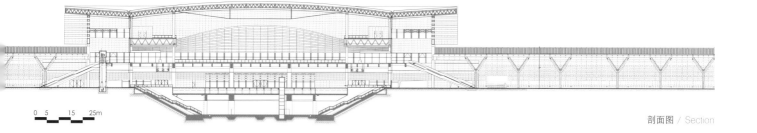

剖面图 / Section

STATION BUILDING OF LANZHOU WEST RAILWAY STATION

兰州西站站房工程

Located in the Qilihe District of Lanzhou City, the central area of Lanzhou, Lanzhou West Railway Station has an area of 221,000 square meters and a gross floor area of 260,000 square meters. It has three floors above ground and one underground, beneath which is a station for Metro Line 2. These two structures were built as an integrated whole. With a maximum elevation of 39.55 meters above ground and 21.26 meters below ground, the station building mainly serves functions of railway passenger delivery and tourism.

01　候车大厅 / Waiting hall

总平面图 / Site plan

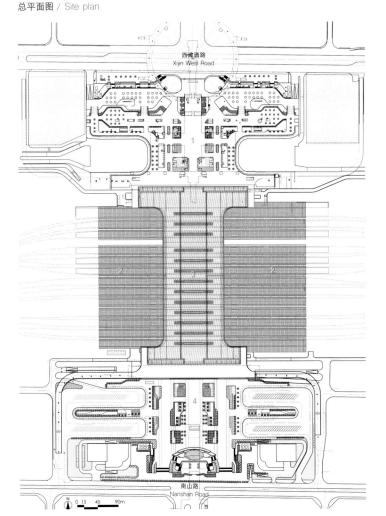

1　北广场 / North plaza
2　雨棚 / Canopy
3　国铁站房 / National Railway Station building
4　南广场 / South Plaza

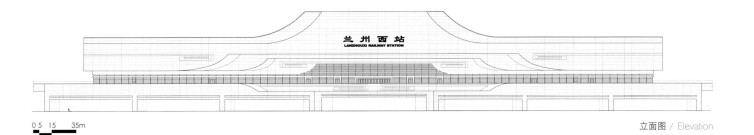

0 5 15 35m

立面图 / Elevation

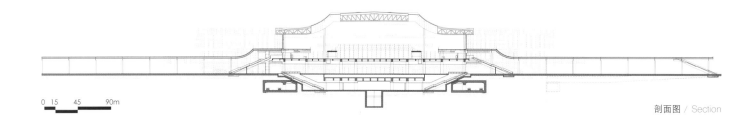

剖面图 / Section

兰州西站位于兰州市七里河区，处在兰州市中心地带。建筑用地面积22.1万平方米，总建筑面积26万平方米，地上三层、地下一层。地下室下方为地铁2号线区间，两者整体建构，地上总高度39.55米，总埋深21.26米。站房主要功能为铁路客运交通，兼具旅客服务功能。

兰州是古丝绸之路上的重镇，曾是历史上著名商贸交通线的重要结点，兰州也是黄河唯一穿城而过的省会城市。为此，建筑外观设计成河流冲过的山体形态，仿佛飘动的丝绸，结合中原和西域的文化风情，展现了兰州西站"黄河丝路"的地域特征和"飞天甘肃"屹然崛起的气势。也充分表达出甘肃人民勇于开拓、再创新"丝绸新路"的时代精神。

兰州西站的室内大空间设计采用室内与室外的一体化的手法，将外部造型的厚重与飘逸手法延续至室内，使内外设计元素高度协调、一致，保证观者感官的一致性。同时，对大厅内部的声环境、光环境也进行了重点研究，如在高架吊顶内设置吸音构造以保证大厅内的混响时间适应旅客的收听要求；在屋面采光处采用遮阳构造，避免产生直射眩光；对高架候车大厅的整个外围的保护体系，也从节能角度进行了加强，如外围防护玻璃幕墙采用双银冲氩气的节能玻璃、地面保温采用的是复合聚氨酯A类保温材料，传热系数低，利于节能，这使得高架候车大厅的候车舒适性得到极大的提高。

02 站台 / Platform
03 外立面透视图 / Exterior facade perspective
04 室内大厅 / Interior hall

一层平面图 / First floor plan

二层平面图 / Second floor plan

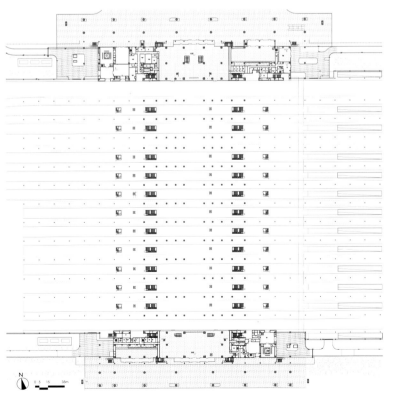
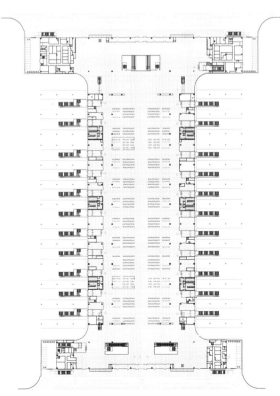

HAIKOU COACH TERMINAL
海口汽车客运总站

Being the supporting facility of Haikou East Station of the Hainan Huandao High-speed Rail Line, Haikou Coach Terminal is the Level 1 coach station of Haikou City, and highlights the operating and distribution functions of the main base. The terminal enhances the connection between the expressway and other transportation modes. Integrating built-in traffic and other comprehensive services, it presents the intensiveness and efficiency of modern city, and offers a provocative urban complex.

总平面图 / Site plan

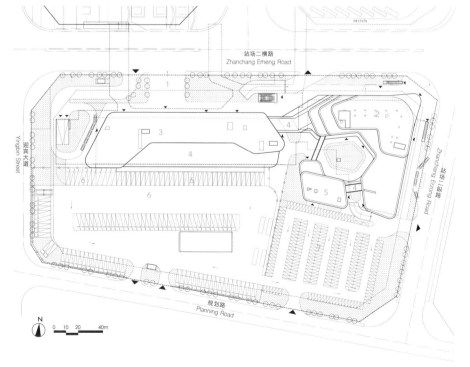

1 站前广场 / Station plaza
2 服务中心大楼 / Service center building
3 站房楼 / Station building
4 室外平台 / Outdoor terrace
5 餐饮楼 / Dining building
6 大巴蓄车位 / Parking lots for buses
7 停车场 / Parking lots

立面图 / Elevation

01 高铁、长途客运站整体鸟瞰图 / Overall aerial view of High Speed Rail Station and Haikou Coach Terminal

02 站房楼人视图 / Perspective station building
03 黄昏局部鸟瞰图 / Aerial view at dusk

海口汽车客运总站是海南环岛高铁——海口东站的配套枢纽，为海口市一级汽车客运站。本项目突出基地的运营功能、集散功能，以提高公路与其他交通方式的对接。置入交通及综合配套服务等多种功能于一体，体现现代化城市的集约与高效，提供一个丰富有趣的城市建筑综合体。

海口汽车客运总站以交通功能为核心，结合区域路网和交通设施的布局明确区域的交通组织，重点分析车站车流的出入交通组织及组织流线。同时进行地块内部交通规划，重点深化站点周边交通接驳研究，构建便捷的立体化换乘体系；并深化站点出入口、停车场（包括机动车及非机动车）等配套设施设计。

考虑海南地域特点和周边环境因素，中心广场上方采用连廊和露台等元素，使室内外两种不同感受的空间互相渗透，为乘客营造舒适的等候场所。通过统一的建筑语汇，鲜明的立面风格，突出交通建筑特有的动感风格。同时车站和服务楼各自相应的空间设计主题，使整个建筑群即分又合、统一而富有张力。

海南琼岛优雅迷人，规划设计从大自然中获取灵感，汲取琼岛的地貌特征，将舒展起伏的岛屿，蜿蜒曲折的海岸线转化为建筑元素予以提炼，规划布局具有鲜明特征。有海就有帆。建筑立面干净简洁，整体造型像流动的音符，更像海岸线附近游弋的洁白风帆，塑造出海口门户的全新形象，寓意着南海明珠的全新起航。

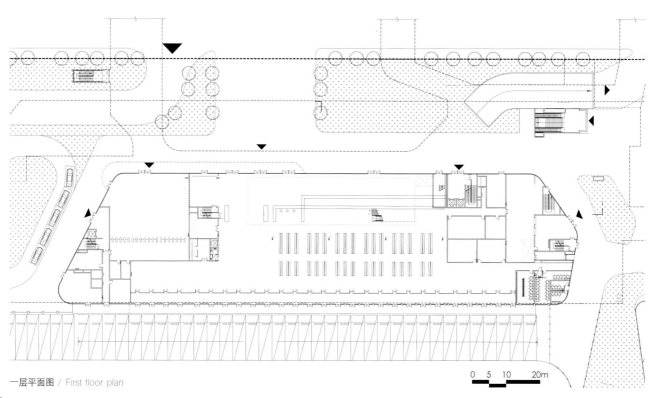

一层平面图 / First floor plan

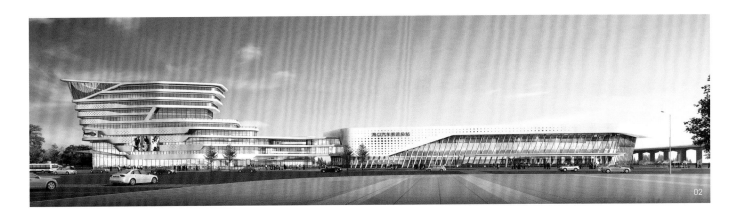

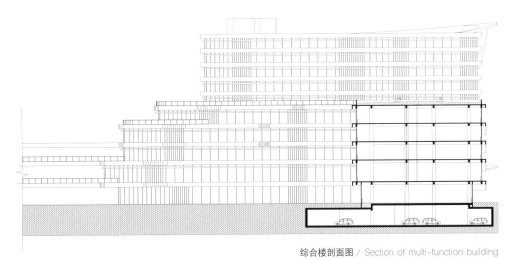

综合楼剖面图 / Section of multi-function building

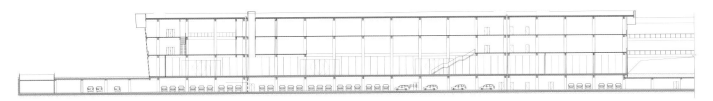

站房剖面图 / Section of station building

FOLLOW-UP PROJECT OF THE SHANGHAI WUSONGKOU INTERNATIONAL CRUISE TERMINAL

上海吴淞口国际邮轮码头后续工程

Located in Wusongkou, Baoshan District, Shanghai, the International Cruise Terminal has a gross floor area of 55,408 square meters. Inspired by the "picturesque scene of the sea," the terminal building echoes the original Oriental Eye in its appearance, forming a continuous whole and creating a magnificent atmosphere for the Wusongkou International Cruise Harbor.

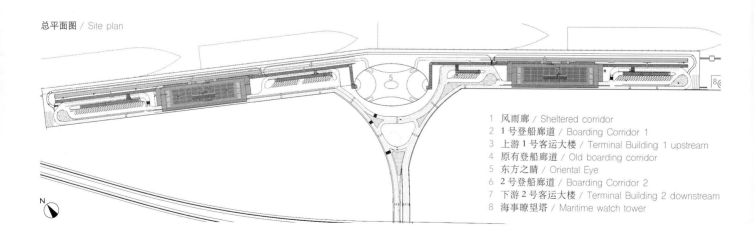

总平面图 / Site plan

1 风雨廊 / Sheltered corridor
2 1号登船廊道 / Boarding Corridor 1
3 上游1号客运大楼 / Terminal Building 1 upstream
4 原有登船廊道 / Old boarding corridor
5 东方之睛 / Oriental Eye
6 2号登船廊道 / Boarding Corridor 2
7 下游2号客运大楼 / Terminal Building 2 downstream
8 海事瞭望塔 / Maritime watch tower

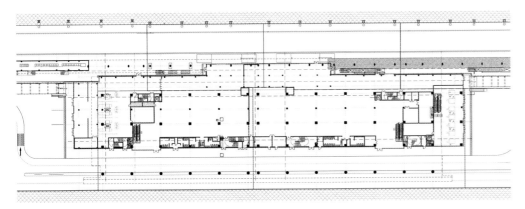

一层平面图 / First floor plan

01 黄昏鸟瞰图 / Aerial view at dusk

221

上海吴淞口国际邮轮码头位于上海市宝山区吴淞口，总建筑面积55408平方米。此客运大楼建筑以"海上画卷"作为意向，来塑造建筑的整体造型，并与原东方之睛建筑形态相呼应，让整个吴淞口国际邮轮港的所有建筑形成一个连续的整体，气势恢宏。

"画卷"既是中国传统艺术的一个重要载体，同时画卷本身还有"蓝图"的意向，寓意着上海城市和上海游轮业的蓬勃发展以及光明的未来。本设计用了山水画的意向来作为画卷的内容，山水本身既代表了东方的文明，也象征着美好的风景，一幅巨大的城市山水画卷展开在江面，与自然相互交融形成一个整体。旅客从这里登船起航，正像是开启一幅美妙的画卷，展开一段美好的旅途。

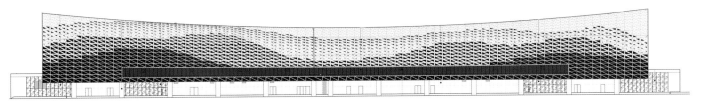

客运大楼立面图 / Elevation of terminal building

02 二层室内边厅人视图 / Interior side hall on the second floor
03 三层办票大厅室内透视图 / Interior perspective of ticket hall on the third floor
04 客运大楼入口侧透视图 / Perspective of the entrance to terminal building
05 瞭望塔日景人视透视图 / Perspective of watch tower

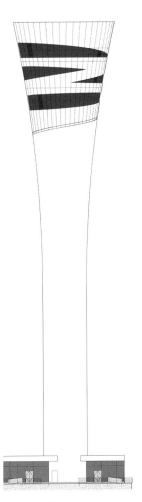

瞭望塔剖面图 / Section of watch tower

LIBRARY IN THE NEW CAMPUS OF BEIJING UNIVERSITY OF CIVIL ENGINEERING AND ARCHITECTURE

北京建筑大学新校区图书馆

Located in Huangcun Satellite City Lucheng Higher Education Park, the library is the core building of the Phase II construction of the new campus of Beijing University of Civil Engineering and Architecture, and the plot takes up an area of 501,000 square meters. It has a ground area of 25,000 square meters and a gross area of 35,000 square meters, including seven floors above ground and one underground.

1 办公入口 / Entrance to office
2 主入口 / Main entrance
3 报告入口 / Entrance to lecture hall

总平面图 / Site plan

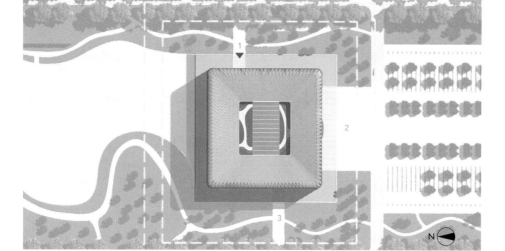

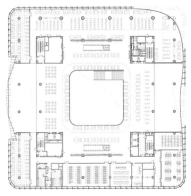
二层平面图 / Second floor plan

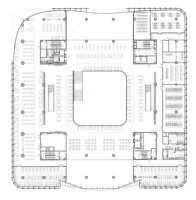
五层平面图 / Fifth floor plan

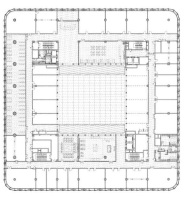
七层平面图 / Seventh floor plan

01 鸟瞰图 / Aerial view

02 北立面外观 / North façade
03 西南侧外景 / View from southwest

北京建筑大学新校区位于黄村卫星城芦城高教园区，占地50.1万平方米。新校区图书馆是二期建设的核心建筑，建筑用地面积2.5万平方米，总建筑面积3.5万平方米，地上七层，地下一层。

北京建筑大学图书馆坐落于校区中央核心景观区之中。以一个69米×69米×30米的几何体作为整个校园最具标志性的符号存在，设计采取高度集中的设计策略来实现图书馆的内在文化承载力度，以此留出宽敞的馆前多层次景观空间作为校园整个学术氛围的延伸与渗透。

建筑的立面以玻璃幕墙和菱形交汇的GRC网格包覆，抽象地对古老而传统的建筑镂空花格窗进行了现代诠释。在连续均质统一的菱形网格模数之中，根据不同朝向的日照及遮阳要求，融入中国传统五行的抽象图解，变幻出富有信息时代特有审美取向特征的立面。

在内部空间的设计上，灵活性与多变性是设计的核心元素。阅览空间自由开放地环绕中庭布置，七层贯通的中庭空间保证了每层都能接受自然光。主题各异的阅览室以一种轻松灵活的方式，螺旋形地插入到不同阅览楼层之中，读者既可通过底层的门厅乘直达电梯到达既定的楼层，也可在环中庭阅览区内随机转换阅览楼层。双重的交通方式恰好与读者在图书馆中的两种空间体验行为——有目的的检索性与无目的的漫游性相对应。如此便为读者提供了一种"无缝"的阅读场所。

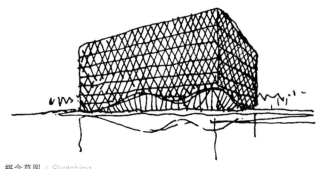
概念草图 / Sketching

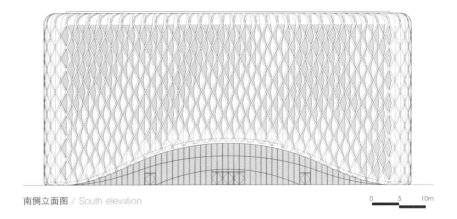

南侧立面图 / South elevation

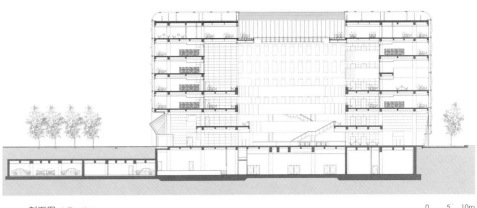

剖面图 / Section

04　入口大厅 / Entrance hall
05　室内局部 / Part of interior view
06　室内回廊 / Interior ambulatory

LIBRARY OF ZHEJIANG COLLEGE OF TONGJI UNIVERSITY

同济大学浙江学院图书馆

Reaching a scale of 30,000 square meters, the library of Zhejiang College of Tongji University will have about 1.4 million books, and will be open to students and teachers of the school. It can accommodate 3,200 readers. Books are offered on the shelves, while meeting halls and research halls in the library allow the organizing of academic events. Occupying an area of about 3,900 square meters, it has one floor underground and ten above ground. It will be an important landmark of the campus after its completion.

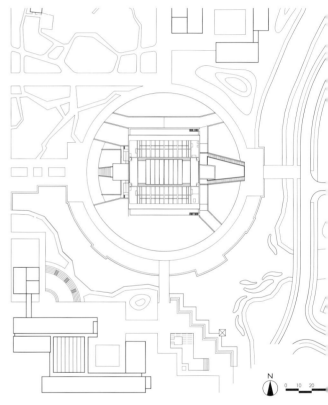

总平面图 / Site plan

01 西侧外观 / West façade
02 鸟瞰图 / Aerial view

03 中庭 / Courtyard
04 接待空间 / Reception area

同济大学浙江学院图书馆的建设规模约30000平方米,设计藏书量约140万册,面向全校师生服务,可容纳人数达3200人。对外功能以开架阅览为主,同时图书馆内部的报告厅、研究室等也可满足多种学术活动。建筑占地面积约3900平米,地下一层,地上十层。图书馆建成后将是校园内的重要地标。

图书馆的方形体量其实是由南北两侧相对独立的两栋板式主楼和它们之间的半室外开放中庭组成的。中庭的底部从地下一层至地上三层横亘着一条由一系列大台阶和绿化坡地组成的往复抬升的景观平台,沿东西方向伸展,将门厅、咨询、大小报告厅、展厅和低层的综合阅览空间等主要公共部分组织在一起。这个半室外中庭是室内外联成一体的,它既能经由西侧架在水池上的主入口通过门厅到达,又可从东侧延伸到河边桥头的室外绿坡直接走到二层平台自由进入。景观平台的上方在东西两侧的不同高度分别设置了数组斜向四边形断面的透明或半透明管状连接体联系南北两侧,内部布置为电子阅览区或会议、接待空间。这使居于建筑内部深处的中庭空间维持了足够的开放度,可以成为校园主轴线上的重要公共空间,它既串联了建筑内外,又在建筑内部提供了依托于中庭体验的多种场所空间。

建筑沿垂直方向分为三大功能区:一至三层及地下一层的公共部分;四至八层全部为复式开架的专题阅览部分;九层、十层分别为研究室和社团活动室,以及带有空中庭院的校部办公室。地下一层在南北两侧与圆形土丘的相接处设置了下沉采光及通风庭院,以改善地下室的气候条件。开放式中庭顶部设有电动开闭屋盖,借由"烟囱效应"有效控制中庭内的空气流动,增进了整个建筑内的自然采光通风,同时保证了中庭内部的气候可控。设计意在通过对开放式中庭和立体景观系统的设置,在建筑中实现一个形态立体化、功能多元化的绿色生态环境和公共交流空间。

东侧立面图 / East elevation

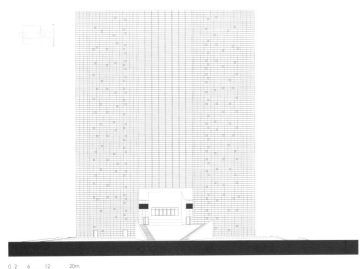

0 2 6 12 20m

剖面图 / Section

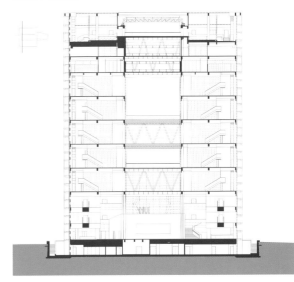

04

二层平面图 / Second floor plan

1 洗手间 / Restroom
2 服务间 / Service room
3 报刊阅览室 / Reading room
4 放映间 / Screening room
5 休息室 / Lounge
6 报告厅 / Lecture hall
7 视听室 / Audiovisual studio
8 电子阅览室 / E-reading room
9 设备管理 / Equipment room
10 空调平台 / Machine platform
11 风机房 / Ventilation

05 专题阅览区 / Reading areas
06 公共空间 / Public space

LIBRARY OF THE CHANG'AN CAMPUS OF NORTHWESTERN POLYTECHNICAL UNIVERSITY

西北工业大学长安校区图书馆

Located in the southwest of Dongda Village, Chang'an District, Xi'an City, the library of the Chang'an Campus of Northwestern Polytechnical University has nine floors above ground, divided into reading areas, meeting halls, an exhibition area, and an office area. It also has one floor underground for equipment and a vehicle garage. The building has an area of 53,631 square meters and a height of 49.5 meters.

总平面图 / Site plan

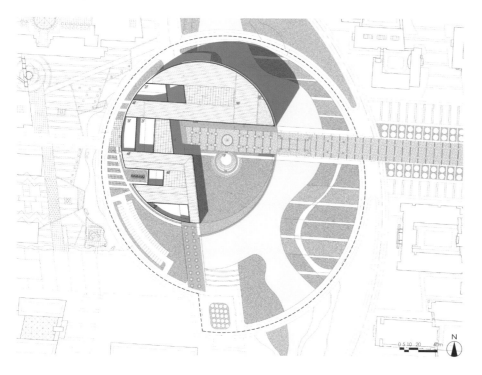

东侧立面图 / East elevation

剖面图 / Section

01　西南方向鸟瞰 / Aerial view from southwest

西北工业大学长安校区图书馆位于西安市长安区东大村西南，图书馆地上九层，按照使用要求分为阅览区、报告厅区、展览区和办公区四部分；地下一层，为设备用房与机动车库。建筑总面积53631平方米，建筑总高为49.5米。

西北工业大学新校区的图书馆，处于校园轴线的交会处——东西向的主入口仪式广场和南北向的景观轴线一贯而过，在校园规划版图中，图书馆所在的这个校园心脏区域是个圆形地块，形式和位置的高度向心性暗示着大学图书馆历来所承载的校园文化精神的使命，决定了这个图书馆的设计目标——创建一个复合各部门功能使用的生理需求与师生们心中知识圣殿心理诉求的"通天塔"。

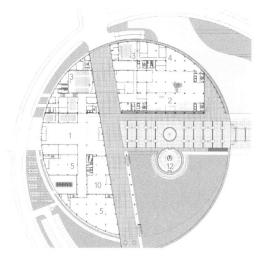 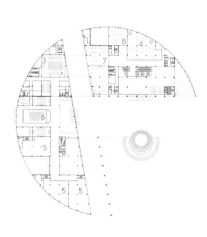

1 中心大厅 / Central hall
2 主入口门厅 / Main entrance hallway
3 报告厅 / Lecture hall
4 阅览空间 / Reading space
5 展览空间 / Exhibition space
6 会议室 / Conference room
7 信息检索中心 / Information retrieval center
8 教室 / Classroom
9 办公室 / Office
10 办公门厅 / Office hallway
11 主题实验室 / Laboratory
12 咖啡厅 / Café
13 中庭 / Courtyard
14 屋顶绿地 / Green roof

一层平面图 / First floor plan　　二层平面图 / Second floor plan

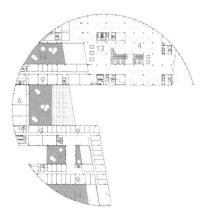 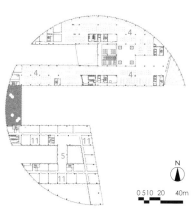

三层平面图 / Third floor plan　　四层平面图 / Fourth floor plan

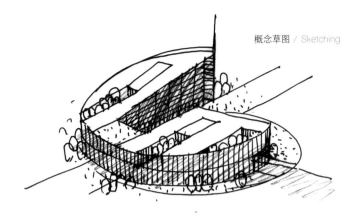

概念草图 / Sketching

02 内街南侧入口 / South entrance at internal street
03 图书馆东北侧外观 / Exterior view from northeast
04 图书馆东侧外观 / Exterior view from east
05 图书馆阅览中庭 / Reading courtyard

JINNAN CAMPUS OF NANKAI UNIVERSITY
南开大学津南校区

The new campus of Nankai University is located in Haihe Education Park in the central urban area of Tianjin and the core area of the golden corridor of Binhai New District. The new campus focuses on emerging disciplines, cross-disciplines and mixed applications, and has reserved a plot for international cooperation, hence has laid the groundwork for building a world-renowned high-level university.

总平面图 / Site plan

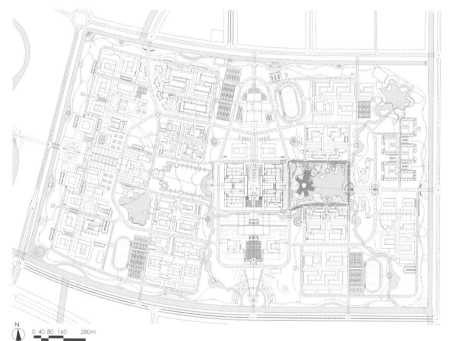

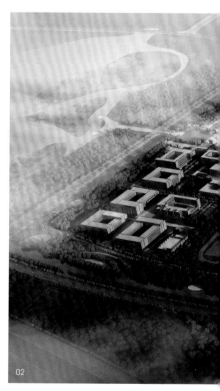

核心教学区底层平面图 / Lower floor plan of core teaching area

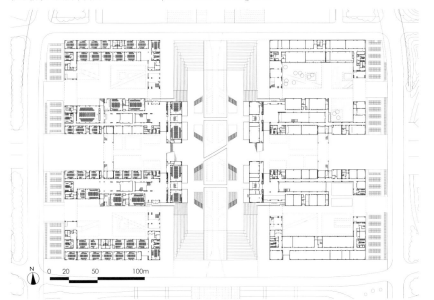

01 核心教学区立面图 / Elevation of core teaching area
02 南开大学日景鸟瞰图 / Aerial view of Nankai University

体育馆东侧立面图 / East elevation of stadium

南开大学新校区位于天津市中心城区和滨海新区黄金走廊的中心位置，海河中游南岸的海河教育园区内。2011年初，新校区的规划开始启动，同济设计集团承担规划设计及大部分的建筑、市政、景观、室内设计工作。

新校区在学科和学院安排上以新兴、交叉、应用为主，同时预留国际合作空间，为学校实现国际知名高水平大学的发展目标留下空间。南开大学新校区总用地面积245.89公顷，规划中校园地上总建筑面积144.35万平方米。

整体校园的规划理念主要为以下四点：强调集聚与共享的学科集群发展，强调人文型、可交流的书院式空间，强调自我调节生长的精明增长模式，强调文脉、建筑与环境共生的校园风貌。

经过演化与具象，最后也就形成了新校区的基本空间特色：

1. 南北资源共享主轴和东西历史文化虚轴交织：资源共享主轴上布置了东西综合业务楼、图书馆、公共教学楼、综合实验楼及体育馆，并预留了发展空间；文化虚轴则以园林广场、纪念建筑群及特色建筑群落串联，寓教于境、润物无声。

2. 三环双心的圈层式组团结构：以中央的东园、西园为两个景观中心，车行主环路、景观步行环路与外圈的复合环路构成三环结构，形成内层教学科研、外层生活运动两个圈层。圈层式结构利于形成便利的交通脉络，内紧外松，各圈层恰好适用于不同的功能类型。

3. 书院式的组群布局：打造教学科研、生活运动的群组群落，包括公共建筑群、院系教学楼群、学生生活楼群、体育中心建筑群、特色建筑群等。最终形成了七个学院组团及其对应的生活运动组团，产、学、研在组团内形成一体化的便利结构，组团间又有着微妙的契合与联系。

4. 可生长式的分期建设：超越传统发展模式，成片预留与就近发展结合应用，满足学科就近扩展与远期学院迁入两种发展需求，打造自我精明增长与生态可持续发展的新生代大学校园。

整个新校区主要包括以下5个组团区域：

1. 核心教学区：位于校区中心，包括公共教学楼及综合实验楼东西两个建筑单体，在建筑设计上以"新书院风格"为核心理念，形成融合东方意蕴与时代精神的"现代书院"特征。

2. 文科学院组团：位于校园的西北角，采用院落递进模式，以四组不同的院落组合，在外部形成统一完整的组群特征，在内部形成空间各异的院落空间。

3. 新兴学科组团：位于校园西南角，组团设计结合校园历史文化轴线及校园内环展开，围绕组团核心绿地紧凑布局，步道与广场收放有致，形成空间丰富、形象特征和谐统一的建筑组群。

4. 体育馆：位于南开大学新校园北侧中轴线末端，建筑平面沿长轴对称，从南到北依次布置综合馆、训练馆和游泳馆，形成三个高低变化的体量。

5. 学生活动中心：位于校园东西文化虚轴中点，临湖而建。建筑包含一个880座的剧院与一个290座的小型音乐厅，还有各类排练用房及学生办公服务设施。建筑设计以南开大学校花"西府海棠"为原型，由六栋大小不等的立方体作放射状排布，围绕中心多边形的内庭园布置交通空间和公共空间，整体造型如花般绽放，营造学生活动中心的活跃氛围。

03 体育馆室内 / Stadium interior
04 体育馆立面 / Stadium facade
05 核心教学区内院 / Internal courtyard of core teaching area
06 核心教学区室内 / Interior of core teaching area

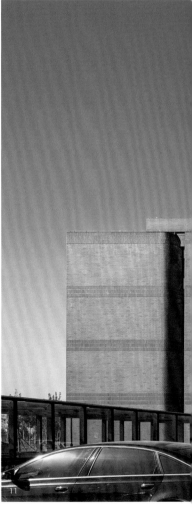
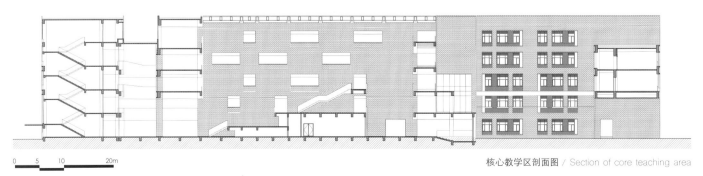

核心教学区剖面图 / Section of core teaching area

07 环境科学与工程学院 / College of Environmental Science and Engineering
08 周恩来政府管理学院 / Zhou Enlai School of Government
09 医学院 / College of Medical
10 哲学院 / College of Philosophy
11 金融学院 / School of Finance

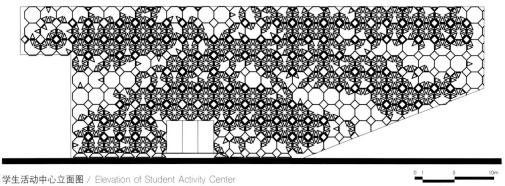

学生活动中心立面图 / Elevation of Student Activity Center

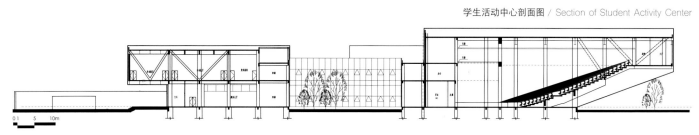

学生活动中心剖面图 / Section of Student Activity Center

12 大学生活动中心鸟瞰图 / Aerial view
13 大学生活动中心室内（一）/ Interior of Student Activity Center 1
14 大学生活动中心室内（二）/ Interior of Student Activity Center 2

1 学生活动室 / Student activity room
2 学生组织办公室 / Office for student organizations
3 视听室 / Audiovisual studio
4 舞台 / Stage
5 室外内庭院 / Outdoor internal courtyard
6 咖啡厅 / Café
7 小音乐厅 / Small concert hall
8 服务大厅 / Service hall
9 合唱团排练厅 / Rehearsal room for chorus
10 交响乐排练厅 / Rehearsal room for orchestra
11 舞蹈排练厅 / Rehearsal room for dance troupe

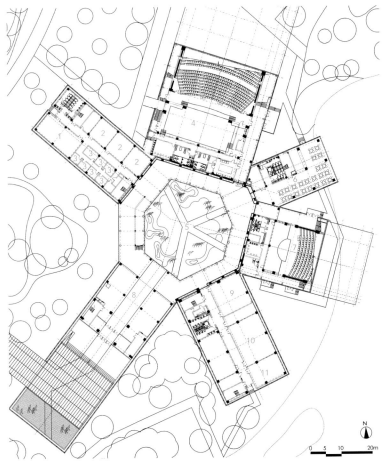

大学生活动中心一层平面图 / First floor plan of Student Activity Center

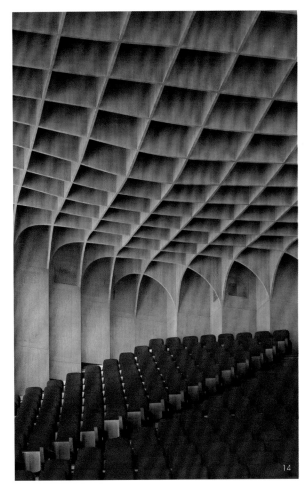

15 大学生活动中心 / Student Activity Center

PUJIANG TOWN JIANGLIU ROAD KINDERGARTEN

浦江镇江柳路幼儿园

As a high-standard educational structure in Pujiang New Town, Jiangliu Road Kindergarten consists of twenty day-care classes and one preschool and faculty training center. Located inside a large, low-density residential area, the west side of the road has been equipped with a main pedestrian entrance and a freight entrance on the north road. The south and east of the kindergarten face residential areas.

总平面图 / Site plan

1　教学楼北栋 / North teaching building
2　教学楼南栋 / South teaching building
3　学前师资培训中心 / Preschool & faculty training center
4　钟楼 / Bell tower
5　机动车入口 / Entrance for vehicles
6　地下车库入口 / Entrance for underground garage
7　教学区主入口 / Main entrance to teaching area
8　培训中心入口 / Entrance to training center
9　园区人行入口 / Pedestrian entrance
10　机动车道 / Driveway
11　公用游戏场地 / Public playgrounds
12　玩沙场地 / Sandpit
13　景观水池 / Landscape pond
14　草坡 / Turf slope
15　绿地 / Greenbelt
16　勇敢者道路 / Courageous Path
17　室外庭院 / Outdoor courtyard
18　停车位 / Parking lots
19　分班活动场地 / Class activity field

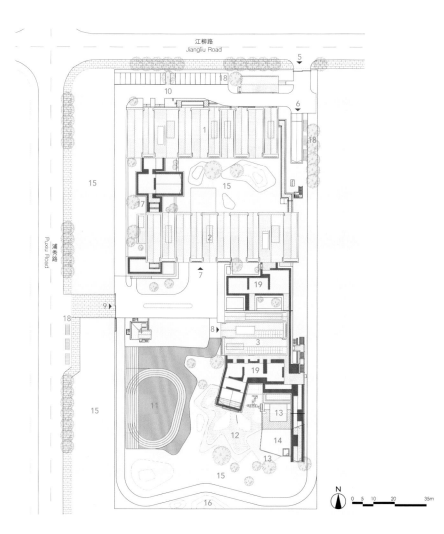

01 从操场看西南外观 / Southwest elevation from playground
02 主入口 / Main entrance

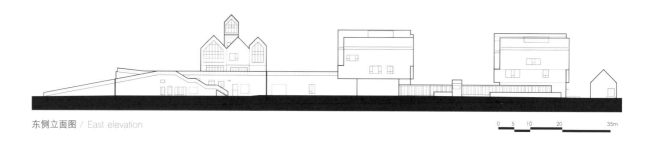

东侧立面图 / East elevation

作为浦江新镇的高标准教育配套项目,江柳路幼儿园由20个日托班和一个早教及师资培训中心组成,位于大片的低密度居住社区内。基地西侧道路设置人行主入口,北侧道路设置后勤入口,东、南两侧与住宅区相邻。

中福会是国内知名的幼儿教育领导者,它对幼儿园设计有自己明确的诉求:一是强调室内外空间的整合关系,创造多层次的幼儿户外活动空间;二是鼓励幼儿的自主成长,将公共空间视为幼儿自主活动的空间载体;三是重视日常运行管理中的安全性与便利性。

本项目的设计就开始于在结合我们的幼儿园设计相关经验的基础上,对于中福会的空间关切的充分回应与引导。总体布局上建筑尽量靠北、靠东布置,留出南侧和西侧大片的户外活动场地。

所有的日托班都在两栋教学楼的二、三层南侧,北侧除了交通、服务设施之外就是一个带有多处放大空间的走廊系统。每个日托班的活动室外都配有可以延展幼儿活动的放大走廊空间,并配有数个贯通上下楼层的小型共享空间,让每个楼层密集的班级空间在这些地方可以得到释放,同时也加强了楼层间的互动。

由于这个幼儿园的规模超越了一般配置,如何控制尺度感知成为设计中的一大重点。在外在形态上,我们将大小差异悬殊的三栋主体建筑都和内部单元空间相对应,同时又有微差的小体量错落叠置而成,并用可以为顶层带来更多空间潜力的双坡顶单元的重复拼接来消解教学楼相对巨大的体量,使主体建筑更接近小房子的抽象聚集。立面上通过不同的开窗方式的交错并置所获得的虚实变化更加强了这种空间意向。

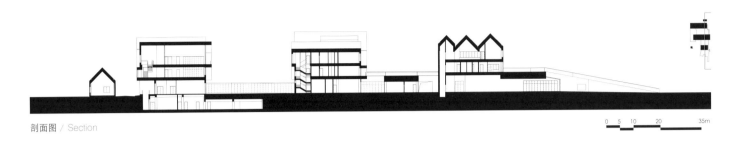

剖面图 / Section

03　主入口 / Main entrance
04　主入口门厅楼梯 / Staircase of main entrance hallway
05　二楼平台 / Second floor platform

06 培训教室 / Training room
07 图书馆 / Library
08 日托活动室 / Day-care activity room

一层平面图 / First floor plan

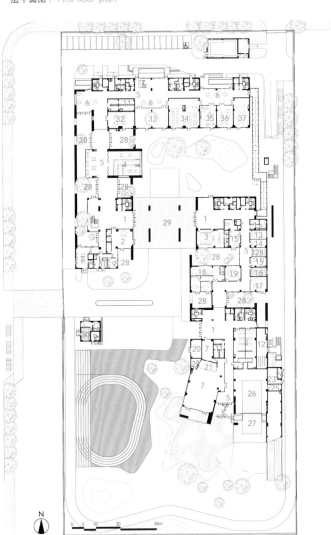

1 主入口门厅 / Main entrance hall
2 晨检室 / Morning check
3 接待室 / Reception room
4 图书馆 / Library
5 公共走廊 / Public corridor
6 室内活动空间 / Indoor activity space
7 多功能厅 / Multi-function hall
8 更衣沐浴区 / Changing room and shower area
9 保健室 / Clinic
10 隔离 / Isolation
11 卫生间 / Restroom
12 热水器间 / Water heater room
12 财务室 / Accounting office
14 人事档案室 / Personnel archives
15 休息室 / Lounge
16 弱电控制室 / Low voltage room
17 科研资料室 / Research reference room
18 办公室 / Office
19 小会议室 / Meeting room
20 活动器材室 / Activity equipment room
21 舞台 / Stage
22 控制室 / CRM control
23 空调机房 / Air conditioning facilities room
24 外机平台 / Machine platform
25 急救室 / Emergency
26 室内游泳池 / Indoor swimming pool
27 室外水池 / Outdoor pool
28 室外庭院 / Outdoor courtyard
29 器械活动场地 / Instrument activity field
30 消防控制室 / Fire control room
31 储藏间 / Storage
32 科学探索室 / Scientific exploration room
33 音乐活动室 / Music room
34 美工活动室 / Painting room
35 生活劳作室 / Life work room
36 认知学习室 / Study room
37 结构木工室 / Woodworking room

THE SCHOOL OF THE SHANGHAI MUNICIPAL COMMUNIST PARTY OF CHINA, PHASE II (TEACHING BUILDING, STUDENT BUILDING)

上海市委党校二期工程（教学楼、学员楼）

The school of the Shanghai Municipal Communist Party of China and Shanghai Administrative College is located at No. 200 Hongcao South Road, Xuhui District, Shanghai. It is a school for the cultivation of mid-level and senior cadres for the city. It must also shoulder the tasks of training senior management for large enterprises and multi-national enterprises in Shanghai, and senior public servants for Shanghai.

总平面图 / Site plan

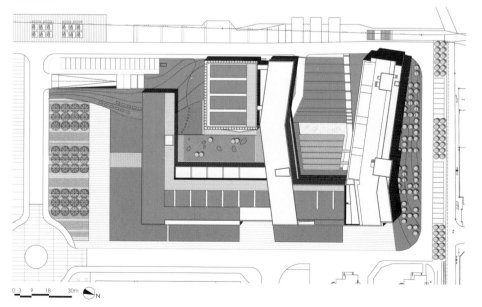

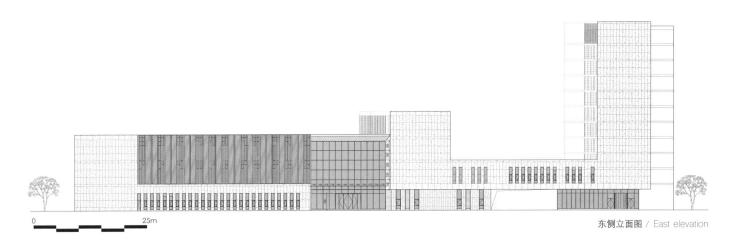

东侧立面图 / East elevation

01 东南面全景 / Panorama of southeast elevation

257

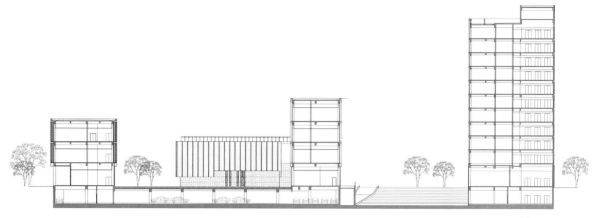

剖面图 / Section

上海市委党校、上海行政学院基地位于上海市徐汇区虹漕南路200号，是一所培养本市中、高级干部的学校，并担负着上海市高级公务员、特大企业及跨国公司在沪机构的高级管理人员的培训任务。

二期工程（教学楼、学员楼）总建筑面积36873平方米，其中地下8426平方米，地上28447平方米。建筑由教学楼和学员楼两部分组成，中间以第二层的连廊相连。教学楼为4层，建筑高度23米，学员楼为11层，建筑高度44.3米。

建筑位于校园西北角，东面和南面为校园景观，西面为漕河泾。建筑东面朝向校园中心绿地，采用平直的界面，山墙面形成两个纯粹的形体："L"和"U"形。西面建筑形体和界面丰富多变，尺度较小，形成几个供人休憩的内院空间。

建筑外立面材料以石材和玻璃幕墙为主。教学楼南面和西面设有竖向遮阳百叶，东面设有大面积种植墙面，报告厅外墙设有爬藤墙面，这些外立面表皮与石材和玻璃幕墙一起形成统一而又富于变化的韵律。室内外空间相互渗透，形成许多饶有趣味的半室外灰空间。

室内设计选用现代的材质和色彩，在部分教室采用了自然光导入技术，大大改善了这些空间的采光。教学楼大厅与屋顶也引入了部分自然光，形成了独特的室内效果。

室外景观与室内空间相互配合，相互映衬。建筑周边安静的水池烘托出建筑静谧的特质，草坡和下沉式庭院是学员休闲和交流的场所。

建筑设计中采用了绿化屋顶、种植墙面、电动外遮阳、自然光导入、地源热泵、雨水回收、智能化集成平台技术、绿色建材等大量绿色节能技术，使本建筑不仅在设计中，也在日后的运营过程中成为一个真正的绿色建筑。

02

02 东南面夜景鸟瞰图 / Aerial night view from southeast
03 下沉庭院日景 / Sunken courtyard

04 教学楼中庭室内 / Courtyard interior of teaching building
05 幕墙遮阳细部 / Details of curtain wall

一层平面图 / First floor plan

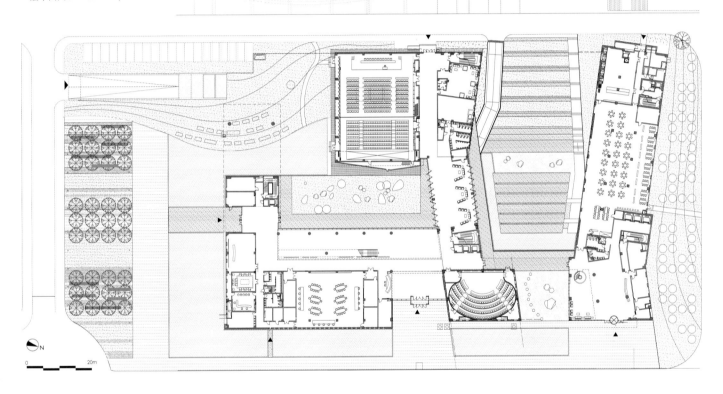

二层平面图 / Second floor plan

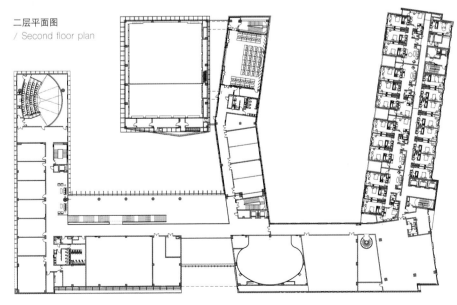

NEW JIANGWAN TOWN ZHONGFUHUI KINDERGARTEN

新江湾城中福会幼儿园

Located in the southwest corner of Plot No. 2 in D Zone of New Jiangwan Town and surrounded by residential buildings, the kindergarten has an area of 14,840 square meters and a floor area of about 8,000 square meters, which will accommodate fifteen day care centers and two early childhood education classes.

总平面图 / Site plan

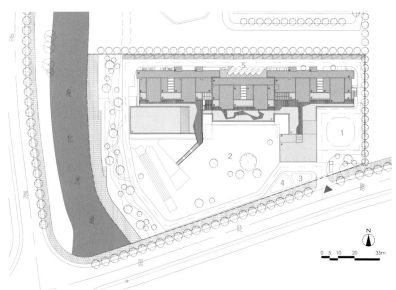

1 广场 / Plaza
2 幼儿活动场地 / Playground
3 种植园 / Planting
4 动物养殖园 / Animal Farm
5 停车场 / Parking lots

南侧立面图 / South elevation

01 南侧外观 / Exterior of south façade

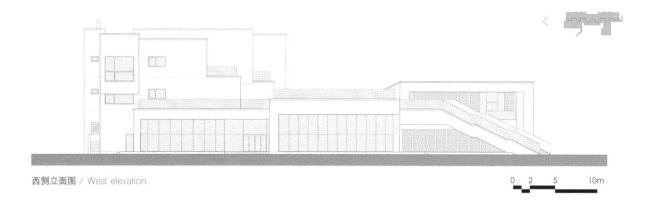

西侧立面图 / West elevation

新江湾城用地位于新江湾城D区2号地块西南角,四周地块以住宅为主。建筑用地面积为14840平米,建设规模控制在8000平米左右,幼儿园建成后可容纳15个日托班及2个亲子早教班的日常使用。

为了回应中福会开放与互动相结合的教学模式,并且充分利用场地条件整合建筑内外环境关系,我们从总体布局到空间规划都探讨了充分适应幼儿身心需要的环境模式。

建筑主体在基地北侧一字排开,留出尽量充足的户外活动场地;三层高的主体分为上下两个耦合的体量,底层是一个容纳了幼儿公共活动及办公、后勤空间的伸出的平台,二、三层退缩的是十五个班的活动室,其中三层的六个班成为微微出挑在平台上的六个盒子。这样的布局方式既为底层的公共活动空间提供了最大的空间弹性及与户外空间联动的可能,又为二层以上的班级活动室提供了巨大的绿色活动平台。

底层的众多公共活动空间由一条东西贯通的宽大中廊相联系,通过庭院以及半室外架空空间的设置,使整个长廊空间有节奏地向南侧的室外空间延展。同时长廊北侧与庭院开口对应的位置还设置了带有天光的垂直的交通空间。长廊中段向南扩展形成一个开放的多功能活动展示空间,形成与周边各专题活动室、大活动室及游泳池相联系的弹性互动空间。以这一长廊为空间主干,将内外、上下串联成为有机的整体。

二、三层的班级活动室都能获得良好的南向采光与南北通风。活动室北面设置宽大的连续走廊,成为班级活动空间的延伸,满足各班级个性化的展示需要。建筑通过退台为每个班级留出了大面积的露台,为幼儿的室外活动提供多种可能。

02 西侧外观 / Exterior of west façade
03 立面局部 / Part of building façade

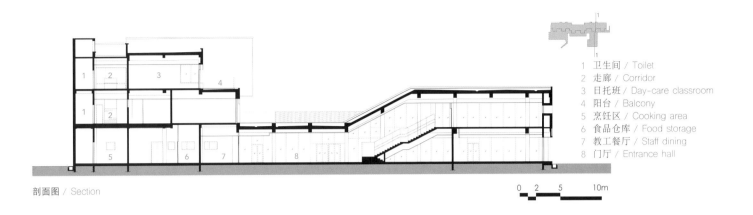

剖面图 / Section

1 卫生间 / Toilet
2 走廊 / Corridor
3 日托班 / Day-care classroom
4 阳台 / Balcony
5 烹饪区 / Cooking area
6 食品仓库 / Food storage
7 教工餐厅 / Staff dining
8 门厅 / Entrance hall

二层平面图 / Second floor plan

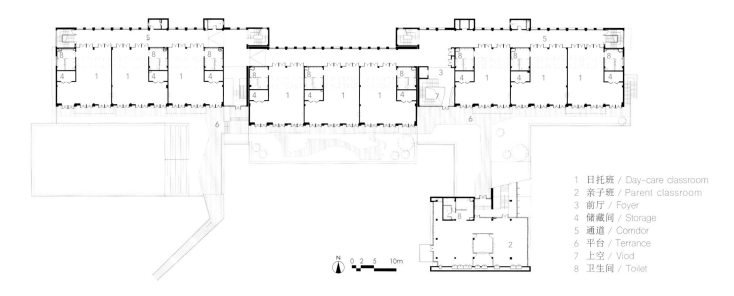

1 日托班 / Day-care classroom
2 亲子班 / Parent classroom
3 前厅 / Foyer
4 储藏间 / Storage
5 通道 / Corridor
6 平台 / Terrance
7 上空 / Viod
8 卫生间 / Toilet

04

05

04 庭院 / Courtyard
05 二层平台（一） / Terrace on the second floor 1
06 二层平台（二） / Terrace on the second floor 2

THE FIRST AFFILIATED HOSPITAL OF SUZHOU UNIVERSITY, PINGJIANG BRANCH

苏州大学附属第一医院平江分院

Pingjiang Branch, the First Hospital Affiliated to Suzhou University, is a Grade III Level-A comprehensive hospital located in Pingjiang New Town in Suzhou City. With a gross floor area of 236,800 square meters, its Phase I takes up 185,974 square meters which is comprised of the outpatient building, emergency building, hospital technology building, inpatient building and boiler house. There are 800 beds in Phase I, which are expected to receive 3,000 cases of outpatient visitors daily.

总平面图 / Site plan

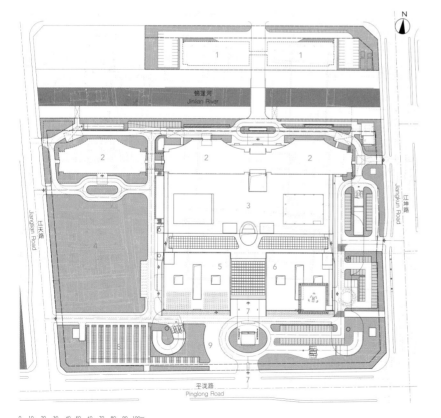

1 行政管理、院内生活楼（二期）/ Administration & living building (Phase II)
2 住院楼 / Inpatient building
3 医疗综合楼 / Medical multi-function building
4 景观绿地 / Landscape greenland
5 门诊楼 / Outpatient building
6 急诊楼 / Emergency building
7 主入口 / Main entrance
8 非机动车停车场 / Parking lots for non-motors
9 广场 / Plaza

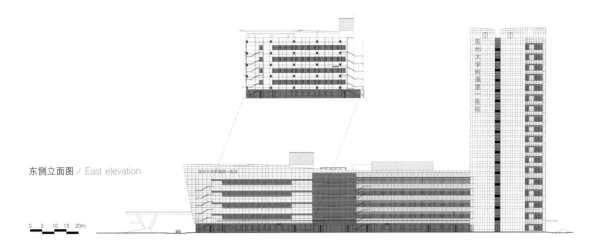

东侧立面图 / East elevation

01　总体透视图 / Overall perspective view

苏州大学附属第一医院平江分院是一家三级甲等的综合医院，基地位于苏州市平江新城。总建筑面积236800平方米，一期工程总建筑面积为185974平方米。由门诊楼、门急诊楼、医技楼、住院楼和锅炉房组成。医院设计一期总床位数为800床，日门诊量3000人次。

基地沿河划分为两大区域，河的北侧为行政管理、院内生活区，河的南侧为门诊、急诊、医技、住院楼。

项目用地较为紧张，将门诊、医技、住院等各个功能紧密的结合起来，为将来的发展确保了充分的空间。远期发展可沿医疗主街方向向两侧用地水平生长，仍可形成高效、集约的整体医疗建筑。

医技楼设置于中心位置，其他功能区域分布在周边。按照部门的关连性将联系紧密的部门集约在同一楼层。门诊楼与医技楼设置医疗大厅，就医目的地一目了然，可提高诊疗效率。

利用门诊楼与医技楼屋顶的较大面积，可通过绿化和雨水回收来降低热负荷，还能提高雨水利用效率。不仅如此，我们还设置了利用地热的地下函道，使自然能源得到有效利用。

在规划中彻底消除步行流线、车行流线、后勤服务流线的交叉，各功能区块分设独立出入口。

基地西南侧留出了大范围的绿化园林区域、广场、绿化与休闲步道有机结合，在医技楼屋面还布置了屋顶花园，其外围也设计有绿化隔离带。

由于道路、河流的限制使整个基地划分成南北两块，为保证建筑的整体性及功能区别，在设计上采取风格相近形态各异的手法，强调其标志性，彰显时代特色与创新精神。

剖面图 / Section

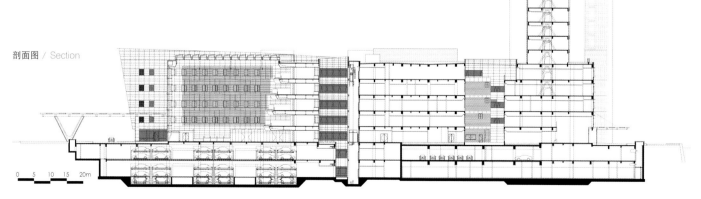

02 夜景透视图 / Night view perspective
03 主入口 / Main entrance
04 门诊大厅 / Outpatient hall

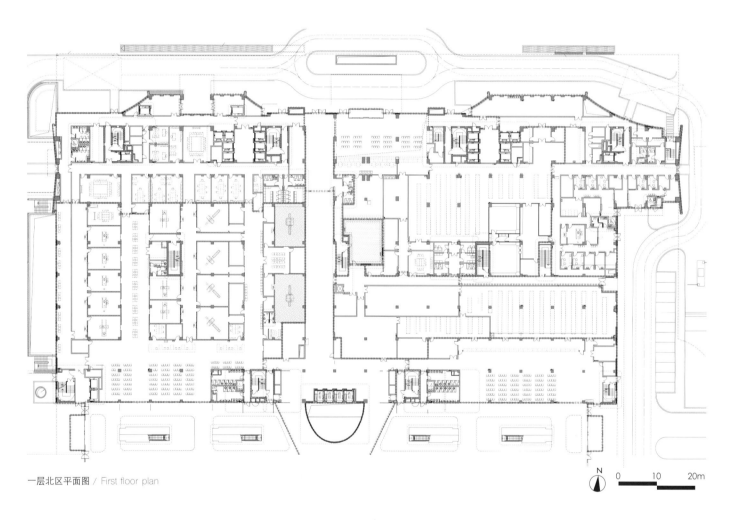

一层北区平面图 / First floor plan

05 会议室 / Meeting room
06 护士站 / Nurse station
07 医疗主街 / Medical passageway

THE NINTH PEOPLE'S HOSPITAL OF SUZHOU

苏州市第九人民医院

Located to the west of Songling Avenue, east of Qiufeng Road, and north of Ludang Road in Taihu New Town, Wujiang District, the Ninth People's Hospital of Suzhou has an area of about 163,245 square meters and a gross floor area of 307,652 square meters, including 224,283 square meters above ground and 83,369 square meters underground. It has 2,000 beds and a designed daily capacity of 7,000 visits.

总平面图 / Site plan

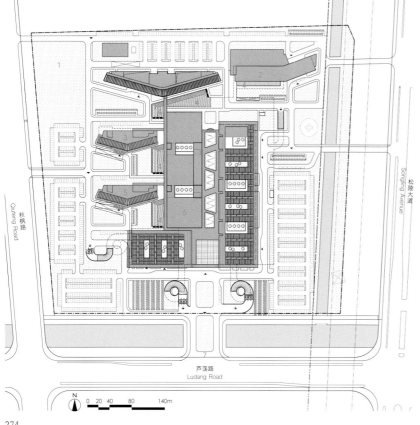

1　预留发展用地 / Reserved site
2　行政后勤楼 / Administration building
3　肿瘤病房楼 / Inpatient building for cancer
4　下沉庭院 / Sunken courtyard
5　妇幼保健楼 / Maternal and children health care building
6　医疗综合楼 / Medical multi-function building
7　主入口 / Main entrance

01 鸟瞰图 / Aerial view
02 西南侧透视图 / Perspective from southwest

03 西侧下沉庭院透视图 / Perspective of sunken courtyard on the west side
04 儿科病房走廊室内透视图 / Corridor interior perspective of children's ward
05 医院街室内透视图 / Interior perspective of hospital passageway

苏州市第九人民医院位于吴江区太湖新城松陵大道以西、秋枫路以东、芦荡路以北。总用地面积约163245平方米，总建筑面积307652平方米，地上建筑面积224283平方米，地下建筑面积83369平方米。医院设计总床位数为2000床，设计日门诊量7000人次。

医疗综合楼采用集中式布局，医技区位于门急诊与病房两者之间，各个功能块之间的距离最短。

在综合医院中设置独立肿瘤专科楼及妇幼保健楼，出入口独立，同时与医院医技门诊部分紧密相连，利用综合医院的医疗资源的同时突出优势医疗学科。

基地内交通组织充分考虑清污、人车分流。主入口前设两个双车道地库出入口。病人通过车库电梯直达医院街，减少地面车流，同时缩短病患步行距离。

院内设公交车站，公交车入院临建筑停靠，病人通过外廊直接抵达门诊大厅，实现公共交通无风雨对接。门诊急诊分别设置出租车蓄客站，方便出租车上下客。

医疗综合楼内结合公共空间设置内庭院，为建筑引入自然通风采光，打造舒适的公共活动及候诊空间。结合医院街设置社会服务站和下沉庭院，以改善病人的就医体验。

诊室设计引入了"医患分开""单人小诊室""分科候诊"等理念，在避免干扰和交叉感染、保护隐私、体现人性化等方面对病人予以充分的关怀，真正做到一切以病人为本。

外立面渐变的水平向构件源自太湖水的意向，主入口处两边向上升起的线条不仅与水的波纹相呼应，同时营造出腾飞的意向。

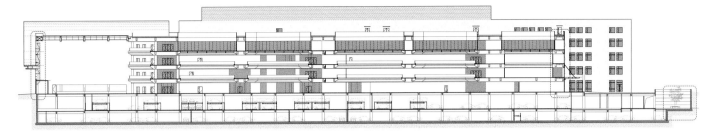

剖面图 / Section

一层组合平面图 / Combined first floor plan

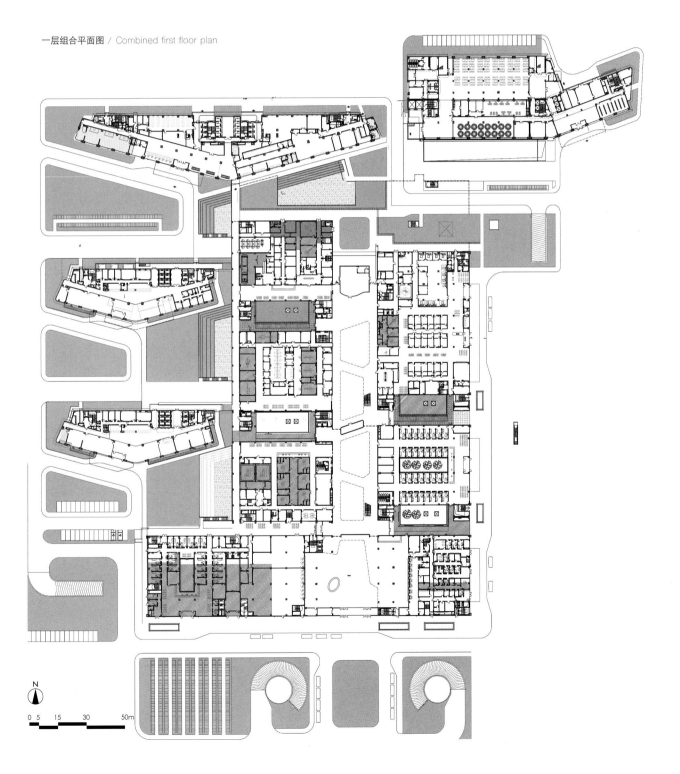

EXPANSION AND RECONSTRUCTION OF THE FIRST PEOPLE'S HOSPITAL OF SHANGHAI
上海市第一人民医院改扩建工程

The expansion and reconstruction project of the First People's Hospital of Shanghai is located at Plot No. 86 on Wujin Road, Hongkou District (formerly Hongkou Senior Middle School), beginning at Jiulong Road in the east and reaching Harbin Road in the north. It is adjacent to a local firefighter control station in the west and Wujin Road in the south. It has an area of about 8,320 square meters (actual measurement prevails) and a gross floor area of 47,904 square meters.

1 中宾楼 / Zhongbin Building
2 医技大楼 / Medical technology building
3 体检中心 / Physical examination center
4 博士楼 / Building for Ph.D
5 硕士楼 / Building for postgraduates
6 门诊楼 / Outpatient building
7 眼科中心 / Ophthalmology center

总平面图 / Site plan

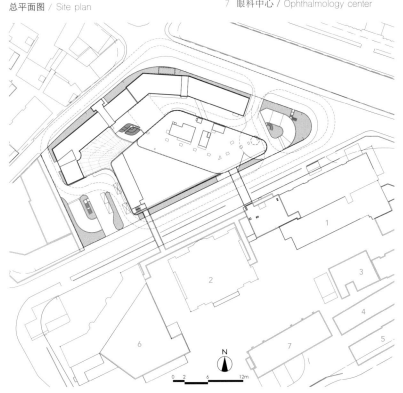

01 鸟瞰图 / Aerial view
02 下沉庭院景观设计 / Landscape of sunken courtyard
03 A楼B楼间的景观设计 / Landscape between Building A and Building B

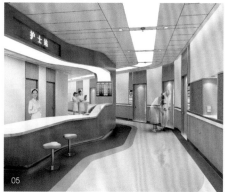
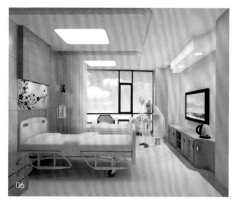

上海市第一人民医院改扩建工程位于虹口区武进路86号地块（原虹口高级中学），东起九龙路、北至哈尔滨路、西侧紧邻市级文保单位消防站，南侧隔武进路与第一人民医院老院区比邻。总用地面积约8320平方米，总建筑面积47904平方米，其中地上建筑面积34340平方米，地下建筑面积13564平方米。高层主楼（A楼）15层，建筑高度61.6米；裙房5层，建筑高度22.4米；保留建筑（B楼）4层，建筑高度16.4米。

本项目将保留原址虹口高级中学教学楼，同时新建一幢具有急诊中心、急救中心、手术中心、中心供应室、功能检查室、病房等功能的综合医疗建筑。同时通过横跨武进路的连廊与武进路南侧老院区形成一个交通便捷、功能互补的整体。

设计总床位数为300，手术室25间，急诊中心设计日均就诊量1000人次。

医院的扩建工程主要遵循了以下设计理念：

1. 绿色医院：采用各种节能减排措施，合理选用建材及设备，降低建筑运行能耗。

2. 人文医院：通过对保留建筑历史风貌的修缮和功能的提升，营造独特的城市中心历史街区的医院形象，倡导人文精神，延续城市历史文脉。

3. 现代化医院：合理布局，充分考虑各功能区与周边环境因素之间的关系，达到资源配置的最优化；建设医院信息系统和综合布线系统，实现就诊的信息化。

4. 人性化医院：营造方便舒适的公共空间，注重近人尺度的细节设计，将人性化的设计贯彻始终。

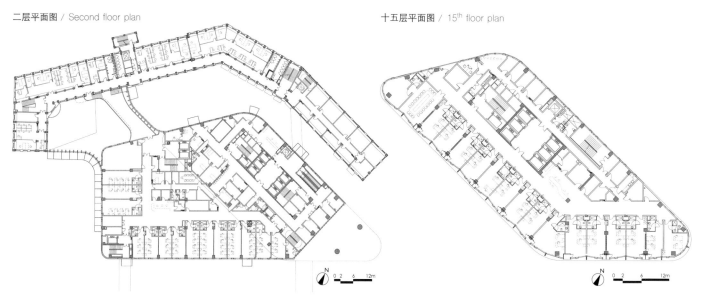

二层平面图 / Second floor plan　　　十五层平面图 / 15th floor plan

04 急诊大厅 / Emergency hall
05 护士站 / Nurse station
06 病房 / Ward
07 沿武进路透视图 / Perspective along Wujin Road

剖面图 / Section

XUHUI DISTRICT SOUTHERN MEDICAL CENTER
徐汇区南部医疗中心

Neighboring Longchuan North Road to the east and Baise Road to the south, Xuhui District Southern Medical Center is mainly aimed at serving Xuhui District Central Hospital (Xuhui Branch of Zhongshan Hospital) and Xuhui Mental Health Center, including three ward buildings and medical ancillary buildings. The north wing and the central building are for Xuhui District Central Hospital, with the north wing used for a surgical ward and the south for the medical ward. The buildings have 16 floors.

总平面图 / Site plan

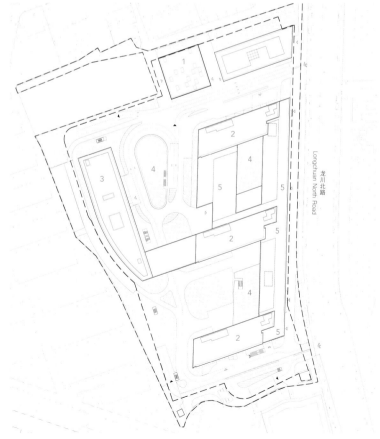

1　体育馆 / Stadium
2　病房楼 / Inpatient building
3　综合楼 / Multi-function building
4　下沉庭院 / Sunken courtyard
5　医疗楼 / Medical building

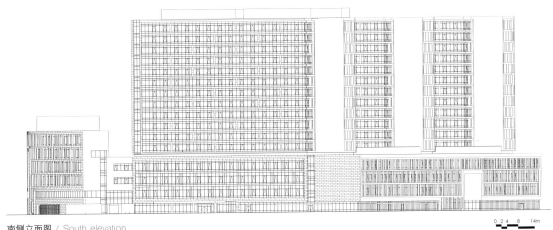

南侧立面图 / South elevation

01　西南鸟瞰图 / Aerial view from southwest

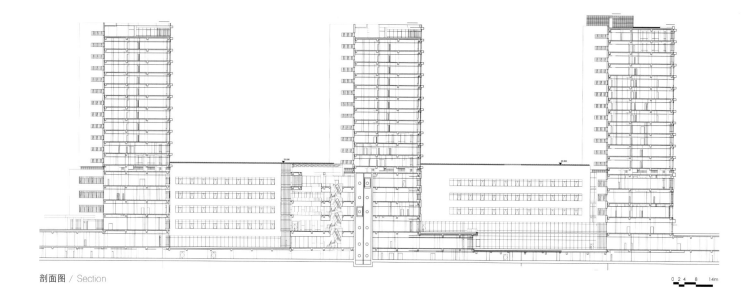

剖面图 / Section

徐汇区南部医疗中心，基地东临龙川北路，南近百色路。项目主体医疗建筑为徐汇区中心医院（中山医院徐汇分院）及徐汇精神卫生中心使用，包括三幢病房楼及相连的医疗楼裙房。其中，北侧和中间两幢病房楼属徐汇区中心医院，北侧为外科病房楼，南侧为内科病房楼，建筑共16层；南侧一幢病房楼属徐汇精神卫生中心，建筑共17层，建筑高度70米。主体医疗建筑裙房属徐汇区中心医院，部分功能为两所医院共用，包括各类门急诊、医技、行政等功能，建筑共4层，建筑高度20.1米。西侧为综合楼，属徐汇区中心医院及中国科学院上海药物研究所，包括各类会议、教育、科研实验、院内生活等功能，建筑共6层、建筑高度30.0米。徐汇区中心医院设计病床950床，手术室15间，介入治疗室3间。徐汇精神卫生中心设计病床500床，不设手术室。

医疗中心的设计秉承了如下设计理念：

1. 朝向植物园的景观面：上海植物园位于用地东侧，设计尝试将三幢病房楼略微转向地块东侧，同时沿东西向均匀错开，使得每幢病房楼南侧与东侧都可以享有朝向植物园的景观面。

2. 将医疗主体建筑置于用地中心：项目周边多为住宅、学校，为了尽可能减少医疗建筑人流、车流、物流对于周边居民的影响，设计将医疗主体建筑置于场地中央。

3. 以绿化和其他建筑单体环绕：尝试以绿化及其他建筑环绕地块北、西、南三侧，以作为周边住宅与医疗主体建筑之间的分隔及缓冲。

4. 由内而外地组织项目基地：项目用地中央的大型广场可将车流迅速引入基地内部，以减少对城市道路的影响，使龙川北路的街道立面更适合步行人流。

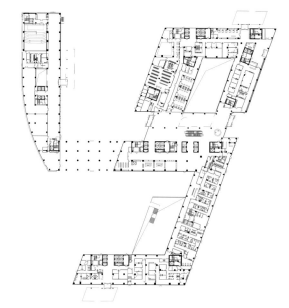

一层平面图 / First floor plan

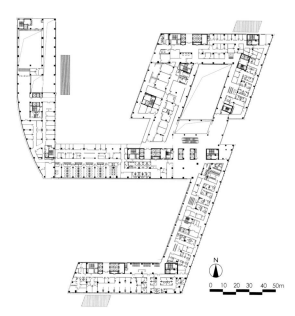

二层平面图 / Second floor plan

02 大厅室内人视图 / Interior of hall
03 综合楼院内人视图 / Internal courtyard of multi-function building
04 东北沿龙川北路人视图 / View from northeast along Longchuan North Road

QINGDAO LINGHAI SPRINGS HOTEL
青岛岭海温泉大酒店

Located by the side of the sea, the hotel resembles a huge ship sailing in the wind, or a bow notched and ready to loose. This shifting appearance is rich in aesthetic tension. The main building is decorated with metal and glass to imply waves, crystallizing a graceful and comfortable atmosphere. The carefully detailed top of the building looks transparent, as if bright lamps shoot forth to light up the night sky.

总平面图 / Site plan

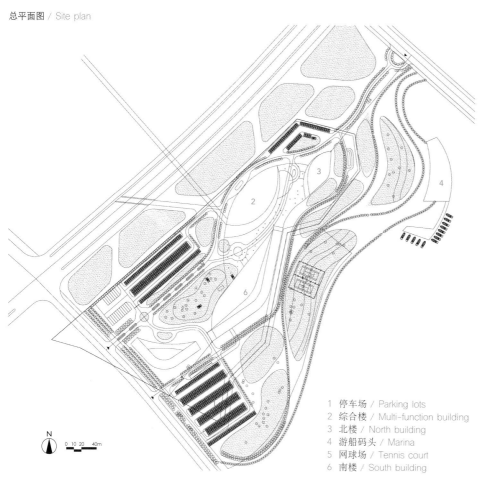

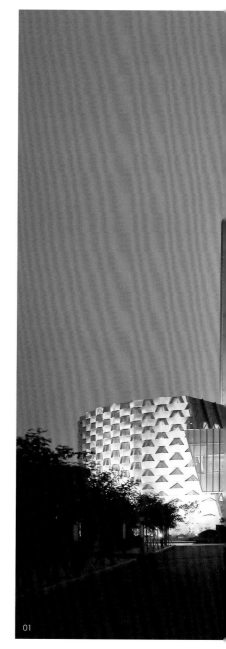

1 停车场 / Parking lots
2 综合楼 / Multi-function building
3 北楼 / North building
4 游船码头 / Marina
5 网球场 / Tennis court
6 南楼 / South building

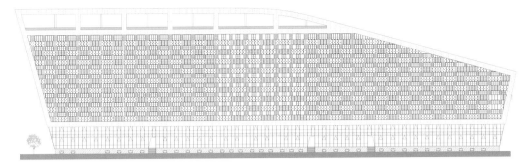

南楼立面图 / South elevation

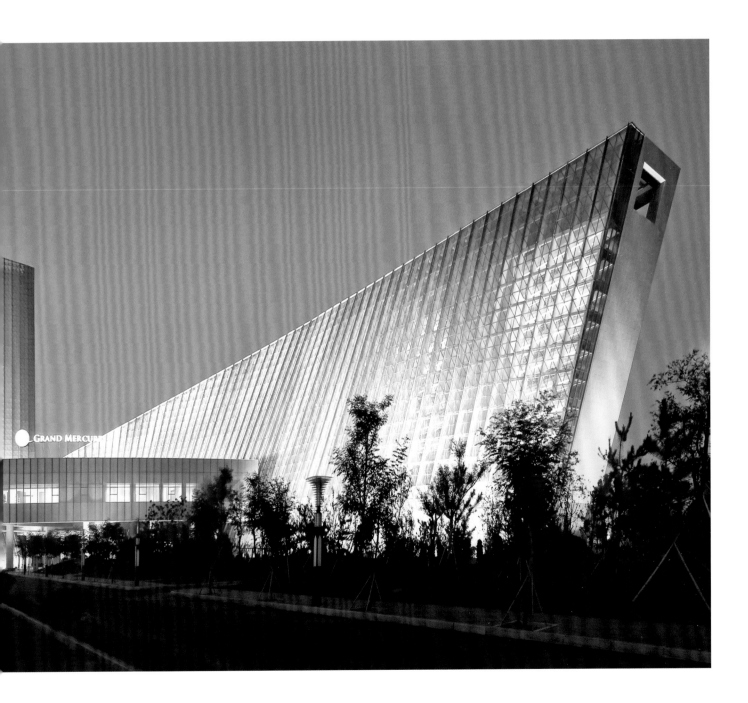

01 夜景 / Night view

02 鸟瞰图 / Aerial view
03 立面肌理 / Façade texture
04 局部透视图 / Perspective
05 入口透视图 / Entrance perspective

项目强化以体验为主导的五星级酒店设计趋向。"环境响应建构"是青岛嶺海酒店设计关注的焦点。环境参数化设计试图提供一个当代建筑实践中的路径与范式。设计由环境参数的理性分析与捕获入手，以热力学能量形式的视角，在风、光、热、景观诸要素的参数作用下，提出了沿海面展开最大景观面，并塑造与风的流动、光的引入响应的如波浪般流畅起伏的总体形态。建筑曲面空间表皮与内部流动空间，成为设计建造的一大挑战。曲面玻璃与斜梁构成了酒店激动人心的峡谷般的中庭空间。

酒店位置紧邻大海，环境优美、视野开阔，给予我们以灵感启示。酒店如同一艘巨轮乘风远航，又如弓箭般蓄势待发，建筑轮廓变化丰富、张弛有致。建筑主体采用金属和玻璃相间的形态，喻义波浪起伏，突出优美舒缓的氛围；精心处理的高层顶部城市之冠通体透明，如同闪耀在城市夜空的明灯。酒店南北楼、绿化庭院、架空平台、屋顶花园、休闲中庭等形成了丰富多彩的连续开放空间系统，室外更通过沙滩和海面而连为一体。

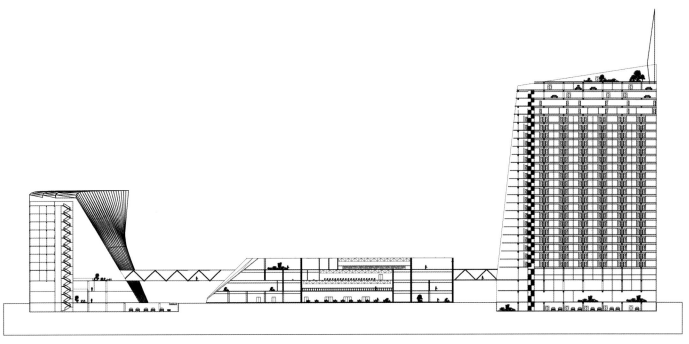

剖面图 / Section

06 中庭透视图（一）／ Courtyard perspective 1
07 中庭透视图（二）／ Courtyard perspective 2

1 公共大厅 / Public hall
2 明档风味餐厅 / Open specialty restaurant
3 休息区 / Resting area
4 韩式料理 / Korean restaurant
5 中餐零点餐区 / Chinese food restaurant
6 海鲜池 / Sea food
7 厨房 / Kitchen
8 包间 / Parlor
9 商务休闲厅 / Business leisure area
10 观景咖啡厅 / Sight-seeing café
11 健身中心 / Fitness center
12 休息大厅 / Resting hall
13 水疗SPA / SPA
14 屋顶花园 / Roof garden
15 商务茶座 / Business teahouse
16 剧院式报告厅 / Theater auditorium

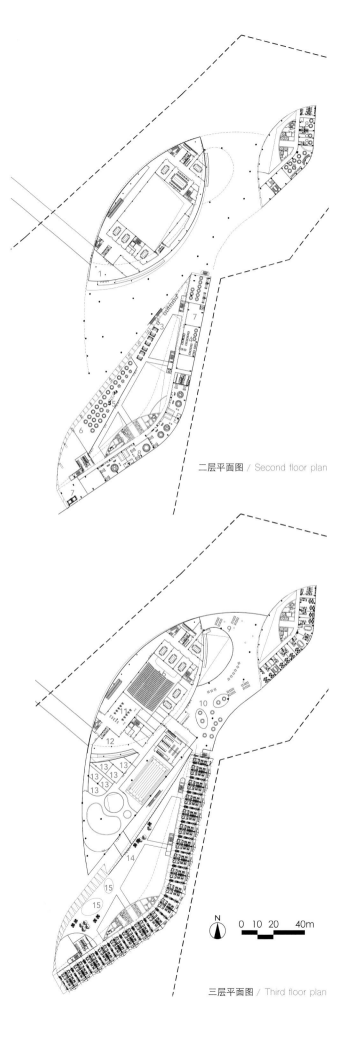

二层平面图 / Second floor plan

三层平面图 / Third floor plan

EXPANSION OF THE SECOND GUEST HOUSE OF THE NATIONAL AFFAIRS MANAGEMENT BUREAU

国家机关事务管理局第二招待所翻扩建工程

Located on the Xizhimen South Street in Xicheng District, Beijing, the site has an area of 24,200 square meters with a gross floor area of 99,980 square meters. The new conference building, with an area of 81,680 square meters, is a complex for meetings and conferences, dining, and accommodation.

01　南立面局部 / Part of south elevation
02　南侧外观 / Exterior view from south

总平面图 / Site plan

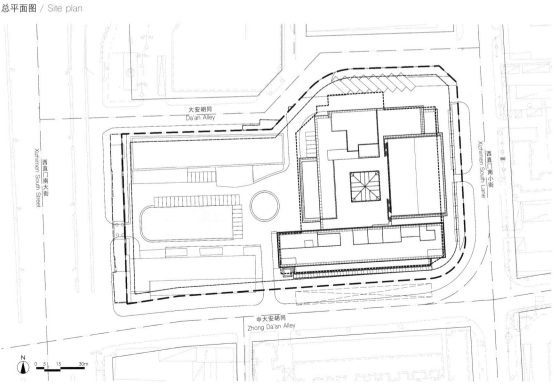

项目位于北京市西城区西直门南大街。建筑基地面积24200平方米，总建筑面积99980平方米，新建会议楼建筑面积81680平方米。新建会议楼是集会议、餐饮、住宿为一体的建筑综合体。地下部分为3层，设置后勤用房、厨房、汽车库及设备用房等功能，建筑面积24702平方米；地上部分南侧塔楼为19层，功能为客房，裙房3层，功能为会议及餐饮，建筑面积为56978平方米。地下三层及地下二层局部设有人防区域。

考虑到该建筑性质的特殊性及所处位置与功能的重要性，在造型设计上，建筑的整体风格典雅、端庄、稳重，以精致的石材为主要材料，在细部精雕细琢，并结合考究的玻璃形成变化细腻的立面感觉。建筑立面采取传统的三段式手法，以厚檐口、凸柱脚、深窗洞的处理方式突出建筑的古典韵味。酒店主楼正立面上当中9跨外凸，体现了在中国传统殿堂建筑正立面"9开间"特征。主界面上以稳重的石材为主，通过两层一组的窗的节奏变化和具有雕塑感的墙面凹凸处理丰富细节，南面两端通过大面玻璃幕墙的处理，使客房层的端套有更好的景观面。主楼顶部结合行政层的功能设置大面落地窗，并通过顶层女儿墙的檐口作为建筑收头。

古典与现代的结合，是这次设计的首要主旨。

03 北侧外观 / Exterior view from north

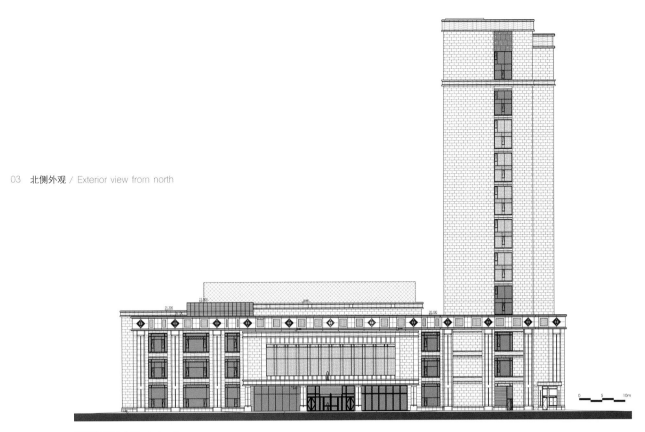

西侧立面图 / West elevation

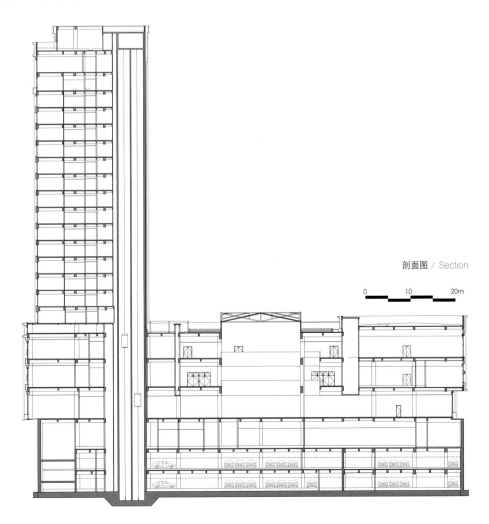

剖面图 / Section

04 仰视图 / Bottom view
05 立面细部 / Details of façade
06 一层入口大厅 / Entrance hall of first floor

一层平面图 /
First floor plan

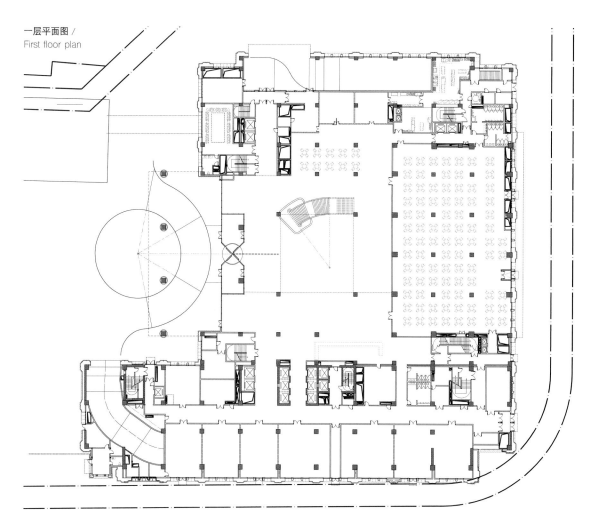

标准层平面图 / Typical floor plan

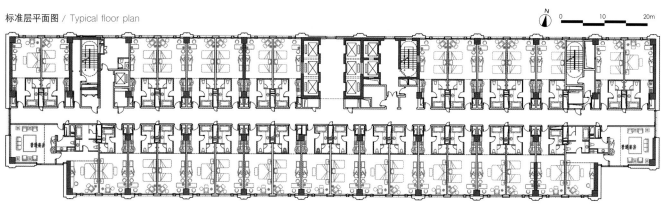

JIMEI MARRIOTT HOTEL

集美万豪酒店

The Jimei Marriott Hotel, located in the central area of Xiting, Jimei New Town, Xiamen City, Fujian Province, has an area of 34,600 square meters. It has 15 floors above ground and 1 floor underground, and has a gross floor area of 99,815 square meters, including 82,700 square meters above ground and 17,115 square meters underground. The building is 65.70 meters tall.

1　酒店入口 / Entrance to hotel
2　主入口 / Main entrance
3　地下车库出入口 / Entrance to underground garage
4　餐饮入口 / Entrance to dining
5　下沉庭院 / Sunken courtyard
6　团体入口、套房入口 / Entrance for groups and suite guests
7　集美新城商务中心 / Jimei New Town Business Center
8　娱乐入口 / Entrance to recreation area
9　地下自行车库入口 / Entrance to underground direct-through garage
10　地面大型停车场 / Parking lots for large vehicles
11　精品店入口 / Entrance to boutique
12　次入口 / Secondary entrance

总平面图 / Site plan

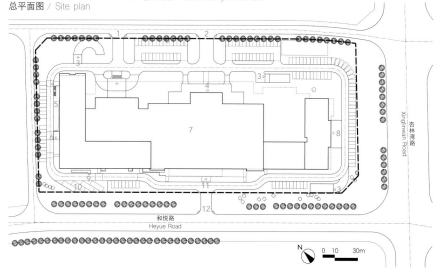

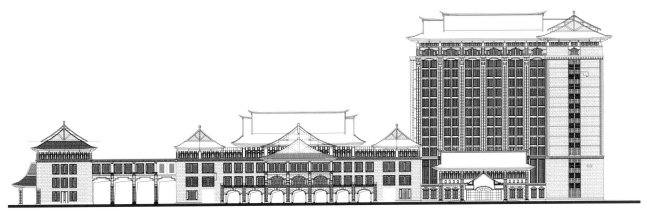

立面图 / Elevation

01 鸟瞰图 / Aerial view

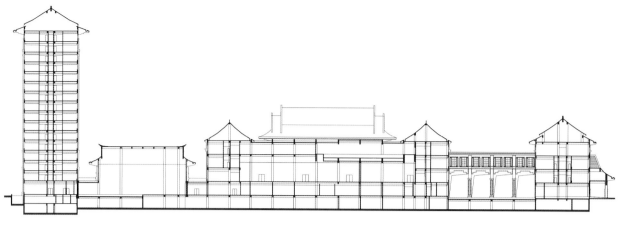

剖面图 / Section

集美万豪酒店工程项目，位于福建省厦门市集美新城西亭中心区。用地面积34600平方米，建筑地上15层，地下1层，总建筑面积99815平方米，其中地上82700平方米，地下17115平方米。建筑总高65.70米。

建筑由主楼和裙房组合而成，地下一层主要是机动车库、设备用房、酒店后勤用房，具体包括餐饮部、客房部，以及工程部，等等，其中机动车库可停车200辆。一层主要是酒店的大堂，以及精品店、KTV、包房。二层居中布置宴会厅、多功能厅、会议室，北侧是酒店公寓，南侧是KTV包房。三层北侧是酒店公寓，其他均为餐饮用房，包括特色料理、中餐厅、大小包厢。四层中部是中餐包厢和活动用房，包括泳池、网球、室内高尔夫等，北侧是酒店公寓，南侧是SPA。北侧客房中心为15层的建筑，呈90度夹角布置，共有客房467个自然间。普通客房自然间面积不小于45平方米。

该设计从城市设计的角度出发，兼顾了本建筑多样化的功能布局特点，做到满足分区明确、交通组织流畅，以及营造优美的环境。建筑造型突出闽南地域建筑文化与建筑形象，完美的演绎了嘉庚建筑风格，在建筑造型与文化之间共同营造建筑、人文、生态、景观和谐共生的环境。

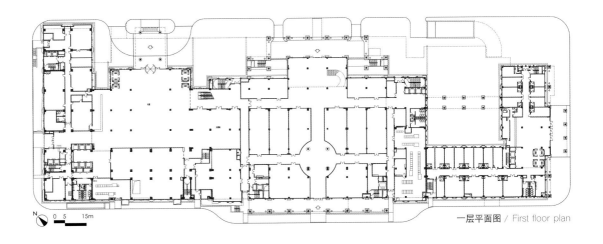

一层平面图 / First floor plan

02 效果图 / Rendering

HOWARD JOHNSON PLAZA HOTEL, FLOWER CITY, KUNMING

昆明花之城豪生国际大酒店

The Flower City of Yunnan is located on Siwa Road, the East Third Ring of Kunming City, and is a large business and office complex integrating a shopping mall, a hotel, and a botanical garden. It has a gross floor area of 250,664 square meters, including 178,416 square meters above ground. The project contains the largest single hotel in China, with 2,268 guest rooms and the largest underground botanical garden in Asia, at over 5,000 square meters.

1 植物资源圃 / Plants nursery
2 西侧主楼 / West main building
3 东侧主楼 / East main building
4 裙房屋顶 / Roof of skirt building
5 主入口 / Main entrance
6 次入口 / Secondary entrance
7 酒店主入口 / Main entrance to hotel
8 商业主入口 / Main entrance to commercial space

总平面图 / Site plan

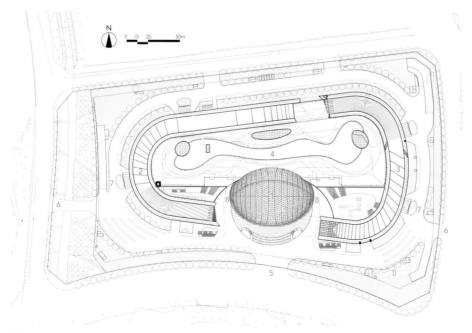

南侧立面图 / South elevation

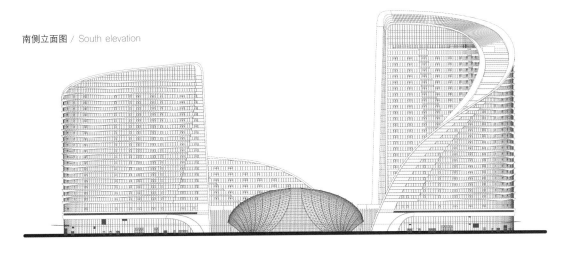

01 湖畔远眺 / Overlook from lakeside

02 植物园室内 / Interior of botanic garden
03 精品商业中庭 / Courtyard of boutique mall
04 植物园室内设计透视图 / Perspective of botanic garden interior

七彩云南花之城项目位于昆明市东三环寺瓦路，是一座集商场、酒店、植物园为一体的大型商业办公综合体。本项目总建筑面积为250664平方米，地上建筑面积为178416平方米。

项目包括了一座超过2268间客房的国内最大单体酒店和超过5000平方米的亚洲最大地下植物园。其中地下停车库可以容纳多达百部以上的大巴客车，位于地下的化妆品旗舰店，其面积超过10000平方米，是世界最大的化妆品单品牌商店。

整个建筑主体形态宛如蓬勃向上的花朵，花瓣绽放，充满生命力，又如一双张开的双手，捧着形如晶莹露珠的植物园。建筑端部层层跌落形如山脉，和周围地势浑然一体，每层退台都是客房层的绝佳观景平台。

建筑主楼立面上水平延续的构件犹如花瓣，通过正弦曲线控制的三维扭转自然形成富于变化的立面肌理和微妙的光感效果。超长尺度的三维扭转水平构件的设计和施工采用数种铝板组件来拟合，在昆明高原的日照下，产生微妙的光影变化，形成充满生命感的韵律节奏。

植物园钢结构主体受力构件的形式源于"叶脉"的原理，钢结构模拟生物自然态，体现自然生命之力形成合理受力模式。在叶脉状的结构体系下，植物园的通风窗、遮阳帘都被巧妙地结合进来。有了这个开敞式的地下植物园，建筑从地下到地面再到裙房、塔楼，形成了真正的立体绿化，创造了充满自然植物的体验。

一层平面图 / First floor plan

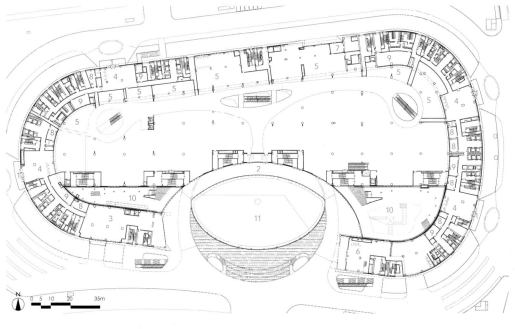

1 酒店主入口 / Main entrance to hotel
2 商业主入口 / Main entrance to commercial space
3 游客服务中心 / Tourist service center
4 酒店大堂 / Hotel lobby
5 商业精品购物中心 / Boutique mall
6 茶室 / Tea house
7 配送中心 / Distribution center
8 办公室 / Office
9 机房 / Machine room
10 室外庭院 / Outdoor courtyard
11 植物园上空 / Overhead of botanic garden

05 主楼内立面局部 / Internal façade of the main building

INTERNATIONAL HOTEL OF HUANGSHANYUANYI BOZHUANG INTERNATIONAL TOURISM EXPERIENCE CENTER

黄山元一柏庄国际旅游体验中心国际酒店

International Hotel of the Huangshanyuanyi Bozhuang International Tourism Center is located to the south of Xin'an River and west of the Peilang River in Tunxi District, Huangshan City, with an area of 69,852.97 square meters. This irregular polygon has a gross floor area of 81,000 square meters with five floors above ground and one underground.

总平面图 / Site plan

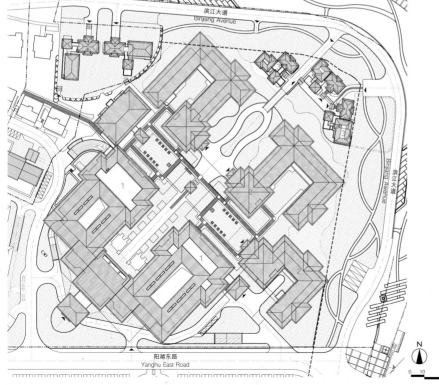

1 酒店主楼 / Main hotel building
2 酒店附楼 / Hotel annex
3 酒店贵宾别墅 / VIP villa

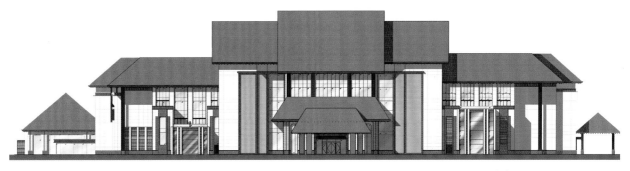

西南立面图 / Southwest elevation

01 鸟瞰图 / Aerial view

黄山元一柏庄国际旅游体验中心国际酒店位于黄山市屯溪区新安江以南，佩琅河以西，总占地面积为 69852.97 平方米，整个场地为一个不规则多边形。总建筑面积 8.1 万平方米，地上五层、地下一层。地下室为酒店后勤及停车库功能，主楼檐口高度 21.95 米，总埋深 7 米。该项目是主要以休闲度假为主要功能的五星级酒店及配套的贵宾别墅。

酒店建筑采用与传统徽派建筑形式相融合的理念。基地中心和南区布置的都是度假酒店，并且成组团布置，建筑整体三至五层。以"U"的平面相互围合形成不同大小的庭院，而庭院正是徽派建筑的一大特点。庭院布置园林景观小品，穿插以石材铺装的游步道，加上有盖连廊能有效地连接各酒店附楼。组团之间也是略微错开布置，一是避免视线干扰，二是形成丰富的墙面空间变化，三是大大提升观赏到江、河两景的客房。沿江的是两层高的酒店别墅，坐拥较优景观。别墅建筑形式采用东南亚建筑元素，开敞通透。

水的引入和利用是设计中的重点和精彩之处，从北面的新安江将水引入基地，再经东侧的佩琅河排出。水面在酒店附楼旁扩展形成一个小的湖面，把江、河两水带到酒店的中心位置，创造一层客房也有观赏江、河两景的机会。

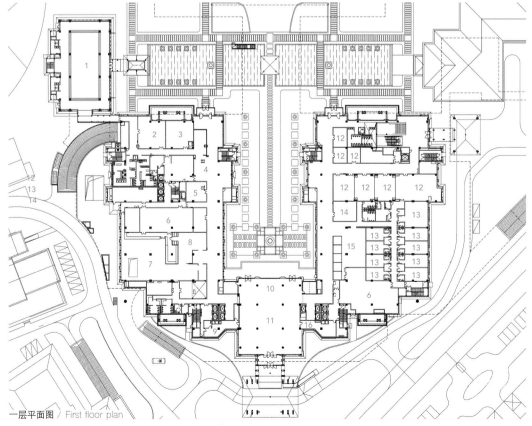

一层平面图 / First floor plan

1 游泳池 / Swimming pool
2 健身中心 / Fitness center
3 儿童室 / Children's room
4 接待厅 / Reception office
5 美容中心 / Beauty center
6 厨房 / Kitchen
7 自助餐厅 / Cafeteria
8 面吧 / Noodle restaurant
9 办公室 / Office
10 大堂酒吧 / Lobby bar
11 酒店入口大堂 / Entrance lobby
12 会议室 / Meeting room
13 贵宾房 / VIP room
14 商务中心 / Business center
15 中餐厅 / Chinese restaurant
16 行李寄存 / Luggage depository

02 客房室内（一）/ Guest room 1
03 客房室内（二）/ Guest room 2
04 屋顶花园 / Roof garden
05 内庭院 / Internal courtyard

RECONSTRUCTION OF THE SHANGHAI HOBNAIL FACTORY (ORIGINAL DESIGN STUDIO)
上海鞋钉厂改建项目（原作设计工作室）

In No. 640 Kunming Road the architects found an old factory built in 1937, then hidden among residential buildings. With the passage of time, it has become an art studio. The reconstruction itself became a sort of archaeological excavation, which dug out the traces of the building's and city's growth, as well uncovering a sense of place and identity for the users.

概念水墨手绘 / Sketching

区位图 / Location map

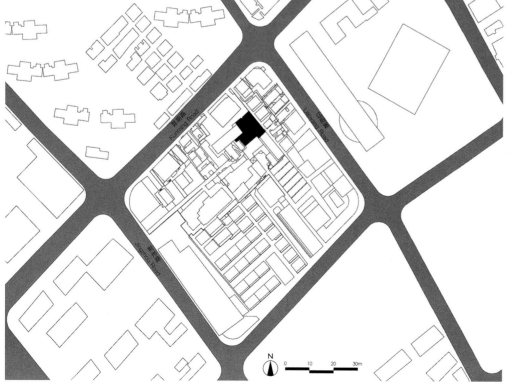

01　原作工作室新址 / New site for Original Design Studio

五院一廊 / Five courtyards and one corridor

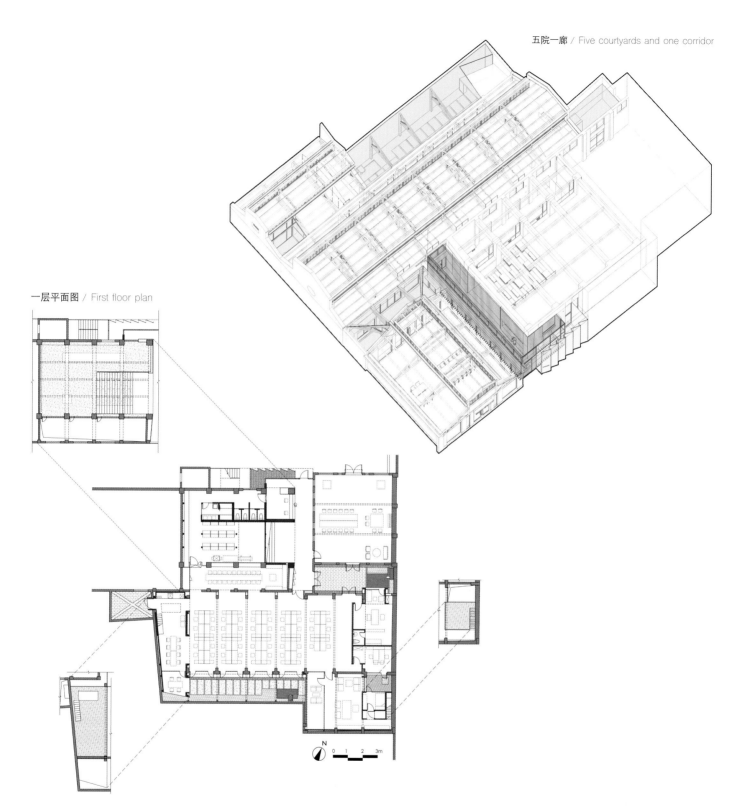

一层平面图 / First floor plan

昆明路640号，隐没在民宅之中的一座始建于1937年的老厂房，历经变迁，从上海鞋钉厂蜕变为原作工作室新址。改造的过程，如同一场考古挖掘，在挖掘出场所沉淀的成长痕迹的同时，也挖掘出使用者的认知定位与本真的内心需求。改造没有刻意抹杀差异的痕迹，使新与旧在改造中占据同样重要的话语权，并更关注新旧对话中张力体系的平衡维系。从老厂房到工作室，建筑类型的变更有利于我们对于空间模式化的规避，异化的空间形成了对空间可能性的重新认知。功能不再成为界定区域的唯一标准，边界的模糊与弥散的体验促进了各种活动的产生。改造的过程也是校正的过程，这种过程状态带来了新的可能，不可预知的问题成为扭转性的启发，使改造过程成为一个允许校正和不断自我平衡的体系。材料不再以既有的模式存在，也不再局限于既有用途，它以更广泛的可能性成就了空间的自由度。

改造暂时完成了对这个充满岁月印痕的场所的应答，因为在1000平方米的空间中，弥散着过去发生的、正在发生的和即将发生的事。

02 人定院 / Rending Courtyard

平面功能图示 / Function zoning

1 展览空间 / Exhibition space
2 研讨、休憩区 / Space for discussion and resting
3 临时工作区 / Temporary working space
4 独立工作区 / Separate working space
5 研讨、展览区 / Space for discussion and exhibition
6 餐饮、休憩、娱乐区 / Space for dining, resting and recreation
7 群体工作区 / Group working space
8 图书、阅览区 / Reading space
9 模型制作、研讨、娱乐区 / Space for modeling, discussion and recreation
10 储藏 / Storage
11 后勤、服务区 / Service space
12 接待、展览区 / Space for reception and exhibition
13 研讨、展览、接待区 / Space for discussion, exhibition and reception

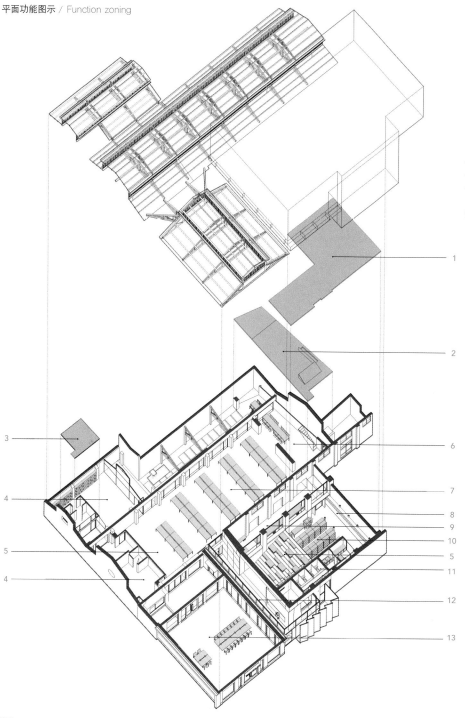

03　隅中院 / Yuzhong Courtyard
04　会议讨论区 / Meeting area
05　工作区全景 / Full view of working space

COLLEGE OF DESIGN AND INNOVATION, TONGJI UNIVERSITY

同济大学设计创意学院

This is a reconstruction and expansion project of a historical building, formerly the site of FAW, which had been divided into a repair factory, workshop, and engineering areas. The design has reconstructed and expanded the original three parts and connected them harmoniously to become teaching, experimental, exhibition, and office rooms of the College of Design and Innovation in Tongji University.

1 原有建筑 (修缮部分) / Old building (Renovated section)
2 设计创意学院新建部分 / College of Design and Innovation (Newly-built section)
3 设计创意学院改建部分 / College Design and Innovation (Reconstructed section)
4 同济科技园 A 楼 (一期) / Building A of Tongji Science and Technology Park (Phase I)
5 同济科技园 (二期) / Tongji Science and Technology Park (Phase II)

总平面图 / Site plan

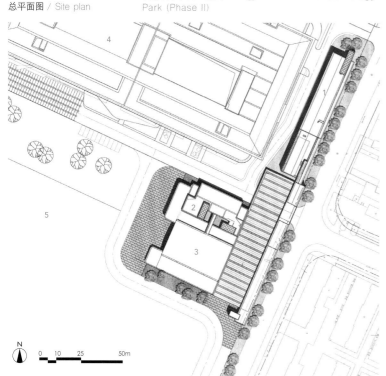

01 外观图 / Exterior view
02 鸟瞰图 / Aerial view
03 入口透视图 / Entrance perspective

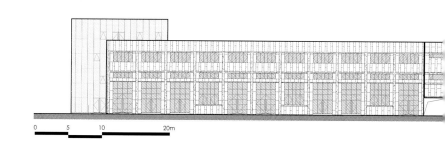

这是一个历史建筑改扩建项目,前身为巴士一汽原厂址,分为修理工厂、流水线车库及工间工程三块。设计将原来的三部分分别进行改建及扩建,并联系起来,成为同济大学设计创意学院的教学、实验、展示、办公空间。改造后的建筑总面积为9007平方米,最高层数为3层。

建筑设计主要采取"嵌入"的策略,保证建筑的原真性、完整性以及可适应性。嵌入主要分为外部嵌入及内部嵌入。外部嵌入主要是通过加建的功能体量,加建垂直交通核,加建的灰空间限定,将原有三个不相关的建筑体量,在水平及垂直关系上联系起来,形成一个建筑整体。嵌入的功能体量在形体、材料、色彩上,与原有建筑一致。内部嵌入主要是在原有的大空间中进行加建,以"盒子"的形式进行嵌入,既保留了原有的空间特性,又化解了原来巨大的尺度,建筑原来是汽车的居所,现在则成为了人的居所。

工艺上以结构加固、材料处理、设备更新为主。结构加固以提高建筑安全性为目的,主要有原柱子外粘型钢、外涂碳纤维、柱间增加钢斜撑、按原比例重做等。材料处理的原则是最大化的保留原有材料,喷涂氟碳保护层,新材料在色彩和质感上,要与旧材料相符,同时保留了大量的线脚——这是原有建筑的立面特征。设备更新则是引入空调系统,加做内保温,保证建筑内部的舒适性。

设计创意学院讲求"互动",也反应在设计上,新旧建筑在空间、形体、色彩、材质无一不反应着这一主旨,一直贯穿在整个设计之中。

04 屋顶平台 / Roof terrace
05 校区入口处过街楼内景 / Interior view of the arcade at entrance

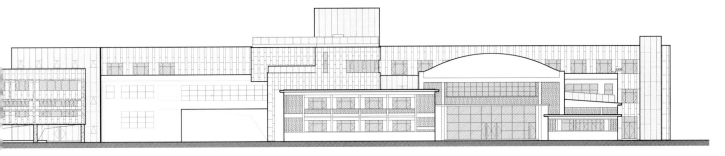

南侧立面图 / South elevation

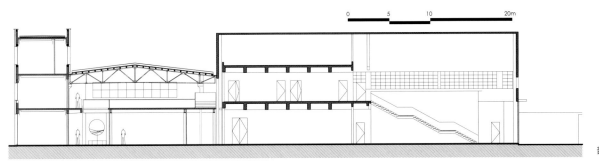

剖面图 / Section

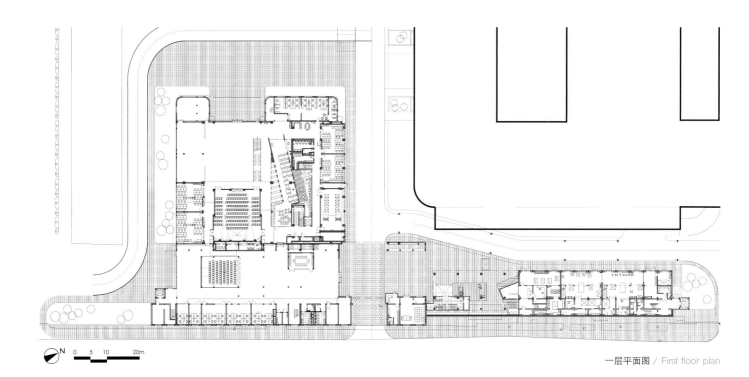

一层平面图 / First floor plan

06　室内透视图（一）/ Interior perspective 1
07　室内透视图（二）/ Interior perspective 2
08　室内透视图（三）/ Interior perspective 3

NO. 816 YAN'AN MIDDLE ROAD (JIEFANG DAILY)

延安中路816号（解放日报社）

The Yan Tongchun Residence was designed by Mr. Lin Ruiji and was completed in 1933. The overall layout of the building suggests it is a Chinese traditional courtyard house with two yards. But the architectural style and the architectural decoration are mostly western, using only some Chinese patterns.

Since 1949, it has served as the office building of Shanghai Instrument Industry Bureau and the headquarter of INESA (Group) Co., Ltd. In 1998, an overhead bridge was built above Yan'An Road, and the road was broadened. The first courtyard of the garden house was dismantled while the main building and the garden were kept as a hotel and guest house. On February 15, 1994, the building was identified as one of the outstanding historical buildings in Shanghai during the second round of nominations.

01　内部庭院 / Internal courtyard

总平面图 / Site plan

A 楼 / Building A
B 楼 / Building B
C 楼 / Building C

1　内庭院 / Internal courtyard
2　景观池 / Landscape pool
3　景观桥 / Landscape bridge
4　下沉庭院 / Sunken courtyard
5　绿化区 / Planting
6　庭院 / Courtyard

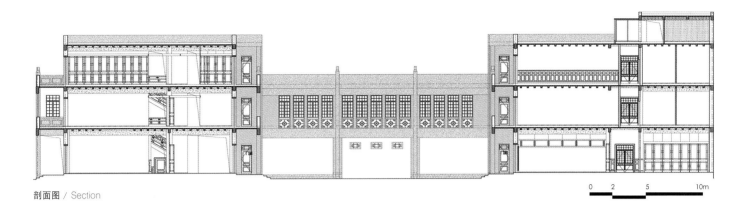

剖面图 / Section

1933年，林瑞骥先生设计的严同春宅落成，建筑总体布局为中国传统的两进四合院型，然而建筑造型和建筑装饰则大都采用西方形式，只是在外观上略加中国图案。

新中国诞生后，先后作为上海市仪表工业局办公楼及上海仪电控股(集团)公司总部。1998年延安路建高架，拓宽马路，该花园住宅第一进被拆除，主楼和花园保存下来，后作为酒店、旅馆等商业用途。1994年2月15日，该建筑被认证为上海第二批优秀历史建筑。

设计旨在通过对历史建筑的保护和解读，更新激活城市空间秩序，充分提升传统街区的当下社会价值及文化内涵。

建筑及场所始建至今，内部院落始终作为核心空间构成元素存在，设计保留核心内院，通过景观整治和立面修缮等方式，强化庭院的主导地位；同时整理院廊空间体系，突出层层递进，景观渗透的空间特质。

设计师在改建过程中尊重了老建筑的真实状态：保留既成的当下真实，又充分保护原始的状态和后续的改动。历史建筑的保护由"原初状态的复原"向"当下真实的保留"转变，历史建筑的现存状态更加反映了当下真实的历时性和即时性。

设计师结合功能的有限介入，使当下的活动参与历史的连续建构。历史建筑不应是历史的见证，通过对主体性的认知，历史建筑可以自觉的介入当下，成为一种自明性的建构过程。最后，设计师应对建筑办公功能的使用诉求，设计提出绿色生态办公模式，通过对保留花园的整理和保留建筑的改造更新，注入新时代、新元素、新需求，叠合传统街区的空间秩序，丰富历史场所的时间积淀。

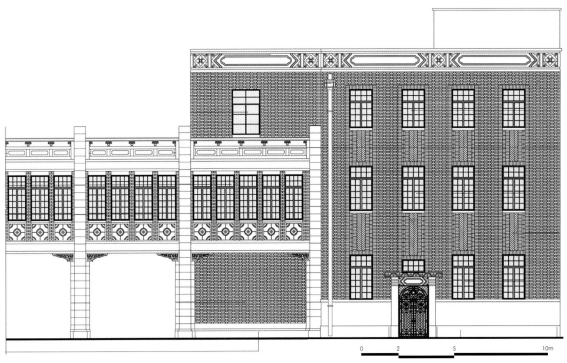

东立面修缮图 / Renovation of east elevation

02　内部庭院 / Internal courtyard
03　保护修缮后的中央庭院 / Central courtyard after renovation
04　B 楼南侧阳台 / South balcony of Building B

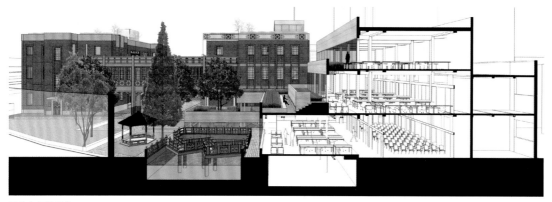

延续空间的秩序 / Continuity of spatial order

05 室内会议室 / Meeting room
06 室内楼梯 / Interior staircase
07 室内走廊 / Interior corridor

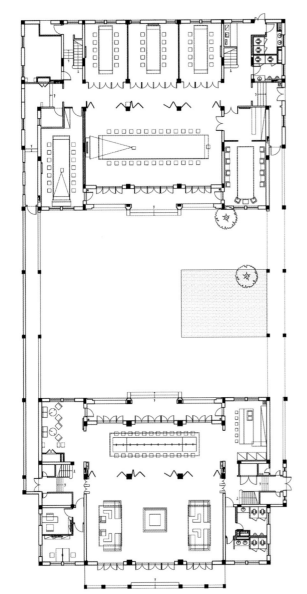

一层平面图 / First floor plan

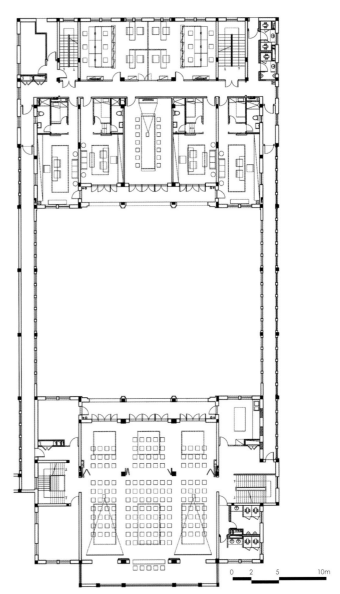

二层平面图 / Second floor plan

MUSEUM OF TONGJI UNIVERSITY
同济大学博物馆

The Museum of Tongji University was reconstructed from Tongji University's 129 Building. With increasing functional management and efficiency and demand to improve the region's level of historical protection, Tongji University set it aside as the school's museum, placed it under protection, and eventually restored its original look. After new functions and equipment were added to the building, it is now capable of receiving visitors and displaying the university's history and culture, and has thus become "Tongji Museum."

总平面图 / Site plan

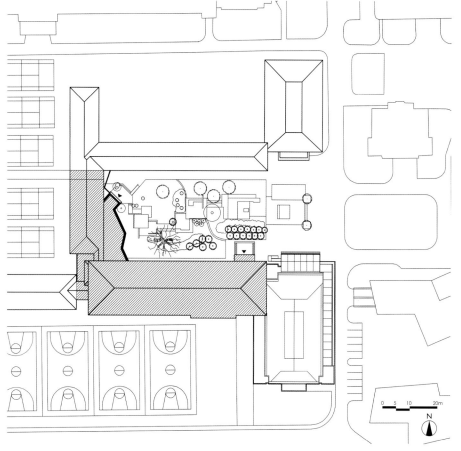

01 改造后博物馆外观 / Exterior view of the reconstructed museum
02 门厅内景 / Interior view of lobby
03 三层展厅内景（一）/ Interior view of exhibition hall on the third floor 1

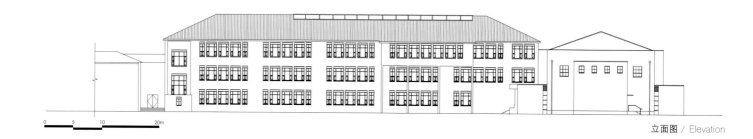

立面图 / Elevation

同济大学博物馆由同济大学一·二九大楼改建而成，该建筑建成于1940年代初，整体呈L形布置，建筑面积4469平方米，地上三层。长期以来，一·二九大楼一直作为学校教学楼使用，随着建筑功能的调整，以及提升区域历史价值的需要，同济大学将其功能定位为校级博物馆。本项目立足于对其进行保护性修缮，恢复历史建筑原有风貌，并在此基础上，通过功能更新、设备更新等技术手段，使大楼在传承历史，延续文脉的同时满足"同济博物馆"的接待展示新功能要求。

改造设计主要从空间价值挖掘、使用功能转换、人文环境融入三个方面进行了深入研究。

一、充分挖掘该历史建筑的价值，如砖木混合的结构形式、早期"日式"建筑的内部功能空间组合关系、相关建筑细部构造等，并反映到博物馆内部的空间设计中。最大化地保留和利用大楼内部的木屋架、木梁结构体系，通过维护和修缮，使其作为结构和装饰构建暴露出来，并在博物馆主要的展示场景中，反映出同济大学悠久的历史。

二、将原有小开间的教学用房改建成适合博物馆需求的收藏和展示空间。同时，增加相配套的功能空间和设备。

三、设计强调对大楼周边环境的整饬和保护，建筑外立面设计严格遵循原有建筑的风貌，如建筑外墙材料、门窗洞口、屋面形式、雨水落管等均按原样修缮。新加建的玻璃门厅，选址在"L"形大楼内转角，即一·二九纪念园背侧隐蔽处。外观设计强调"新旧对比"和"通透性"原则，通过将门厅设计为不规则折线形的通透玻璃厅，一方面避让纪念园的多棵古树，另一方面弱化门厅的形体和体量，使得新增设施对纪念园和原有老建筑的影响降至最小。

剖面图 / Section

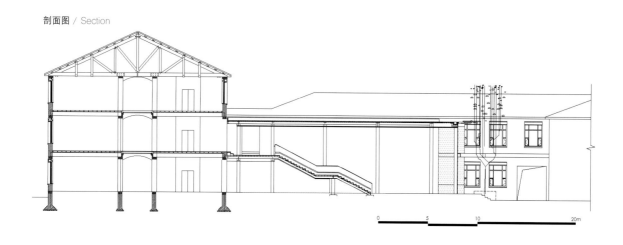

04 从"一·二九纪念园"看新建门厅 / View of the new lobby from 129 Memorial Garden
05 博物馆主入口 / Main entrance to museum

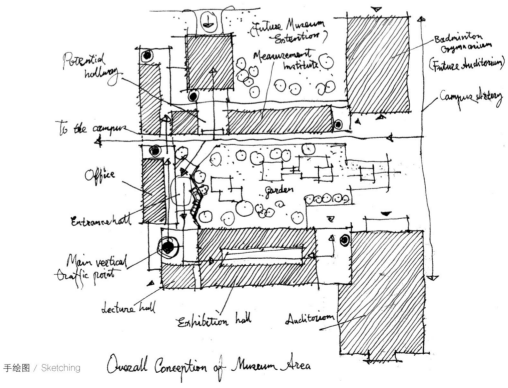

手绘图 / Sketching　　Overall Conception of Museum Area

06 三层展厅内景（二） / Interior view of exhibition hall on the third floor 2

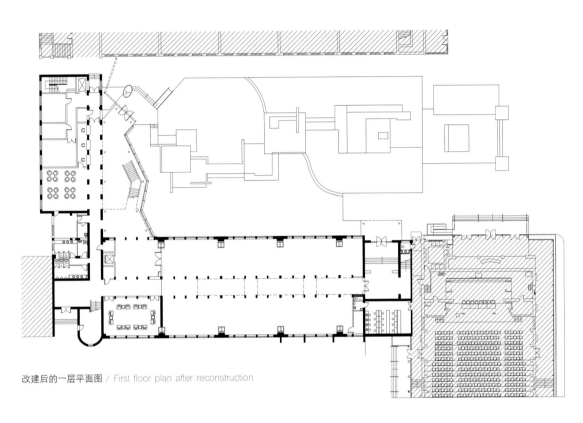

改建后的一层平面图 / First floor plan after reconstruction

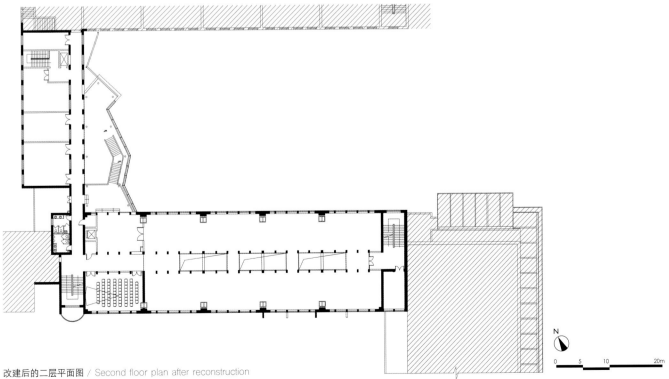

改建后的二层平面图 / Second floor plan after reconstruction

RECONSTRUCTION OF BUILDING D OF THE COLLEGE OF ARCHITECTURE AND URBAN PLANNING, TONGJI UNIVERSITY

同济大学建筑与城规学院 D 楼改建项目

Formerly known as the "Energy Building" of Tongji University, Building D of the College of Architecture and Urban Planning was built in 1978. It is a five-floor shear wall structured building with pre-cast beams and cast-in-situ concrete columns. The building presented a new spatial pattern after its reconstruction, which now integrates Building A, B, and C into a unified teaching block to optimize resource allocation and the teaching environment of the College of Architecture and Urban Planning.

01 D 楼外景 / Exterior view of Building D

总平面图 / Site plan

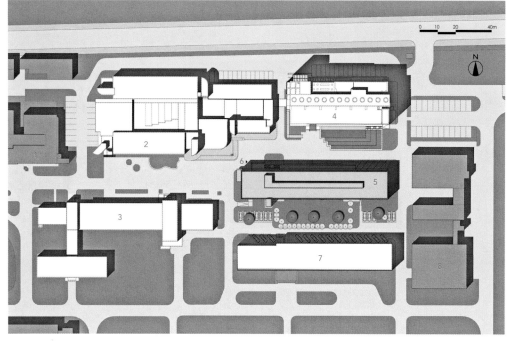

1 留学生楼 / Overseas student building
2 建筑城规 B 楼（明成楼）/ Building B (Mingcheng Building)
3 儿建筑城规 A 楼（文远楼）/ Building A (Wenyuan Building)
4 建筑城规 C 楼 / Building C
5 建筑城规 D 楼 / Building D
6 主入口 / Main entrance
7 外语楼 / Building of School of Foreign Languages
8 同济设计院 / Tongji Architectural Design Group

同济大学建筑与城规学院 D 楼改造前为同济大学"能源楼",该建筑建于 1978 年,为五层框架剪力墙结构、预制梁与现浇混凝土柱整体装配式结构。由于该建筑建造年代较早,建筑材料出现不同程度的劣化,且原有功能布局已经无法满足建筑与城市规划学院新的教学需求,因此对其进行生态节能改造,使其以全新的空间组织和形态与建筑城规学院现有的 A、B、C 三栋楼形成整体学院教学街区,从而优化建筑城规学院的资源配置和教学环境。

同济大学建筑城规学院 D 楼位于建筑城规学院 C 楼南侧,文远楼东侧。其与建筑城规学院 A、B、C 楼形成一定的围合关系,表现为两个既分割又联系的长方形广场。

D 楼强调在保留原建筑结构和形式的基础上,对建筑形态和内部功能进行更新。首先,将原有小开间的办公型单一教学研究用房改建成适合城规学院要求的开放型大空间,集教学、实验、会议功能于一体的复合型教学设施。其次,建筑二层实验室北侧加建钢结构平台,增强了 D 楼北侧界面的公共性,并与 C 楼入口的下沉花园和灰空间形成了良好的呼应关系。再次,采用创新的设计手法,将遮阳装置融合到建筑立面设计之中,形成具有功能与装饰双重作用的外围保护体系。建筑南北侧立面通过钢龙骨挑出活动式冲孔铝遮阳板,可以手动控制遮阳板效果,满足不同季节、不同时间的遮阳和采光要求,造型上也形成了有开有合,统一而富有变化的效果,符合教学楼的气质特征。

02　D 楼外景 / Exterior of Building D
03　D 楼折叠式金属活动外墙板 / Movable folding cladding panels of Building D
04　五层国际交流内廊 / Internal corridor of International Exchange Center on the fifth floor
05　底层模型展示间 / Models showroom on the first floor
06　底层门厅 / Lobby on the first floor

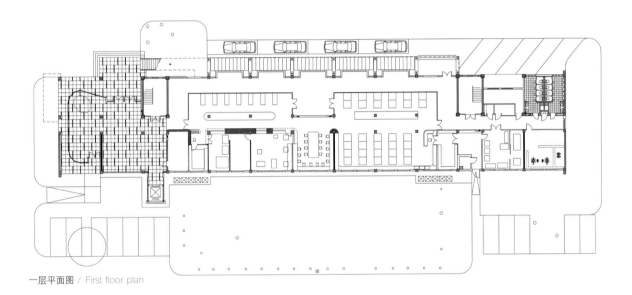

一层平面图 / First floor plan

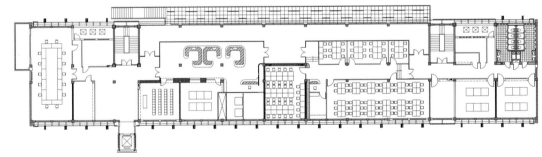

二层平面图 / Second floor plan

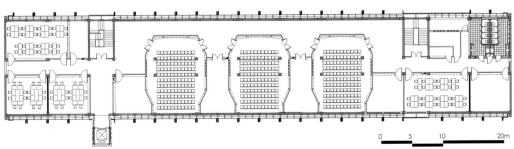

五层平面图 / Fifth floor plan

DATA OF SELECTED WORKS
附录：作品数据

上海中心大厦

项目地点 / 上海市浦东新区
合作设计单位 / Gensler, Consentini, Thornton Tomasetti
规模 / 地上 407570 m²; 地下 165653m²
业主 / 上海中心大厦建设发展有限公司
建成时间 / 2015 年
主要用途 / 办公、会展、酒店、观光娱乐、商业等
获奖情况 / 2016 中国高层建筑创新奖

Shanghai Tower

Location / Pudong New District, Shanghai
Collaborators / Gensler, Consentini, Thornton Tomasetti
Floor area / Ground 407,570 m²; Underground 165,653 m²
Client / Shanghai Tower Construction and Development Co., Ltd.
Completion / 2015
Function / Office, convention and exhibition, hotel, entertainment, commercial use
Award / Innovation Award from Chinese High-Rise Building Award in 2016

上海自然博物馆

项目地点 / 上海市静安区
合作设计单位 / Perkins+Will
规模 / 地上 12128 m²; 地下 32958 m²
业主 / 上海科技馆
建成时间 / 2013 年
主要用途 / 博物馆展览
获奖情况 / 2015 年上海市优秀工程勘察设计一等奖

Shanghai Natural History Museum

Location / Jing'an District, Shanghai
Collaborators / Perkins+Will
Floor area / Ground 12,128 m²; Underground 32,958 m²
Client / Shanghai Science and Technology Museum
Completion / 2013
Function / Exhibition
Award / First Prize from the Shanghai Award of Excellent Engineering, Survey, and Design Industry in 2015

北川地震纪念馆

项目地点 / 四川省绵阳市
规模 / 14280 m²
业主 / 绵阳市唐家山堰塞湖治理暨北川老县城保护工作指挥部
建成时间 / 2013
主要用途 / 纪念、展示、科普、教育、科研
获奖情况 / 2015 年 ARCASIA 亚洲建筑师协会建筑奖金奖

Beichuan Earthquake Memorial Museum

Location / Mianyang, Sichuan Province
Floor area / 14,280 m²
Client / Headquarters of Tangjiashan Yansaihu Lake Control and Beichuan Old Town Conservation, Mianyang City
Completion / 2013
Function / Memorial, display, science, education, scientific research
Award / Gold Medal from the ARCASIA Awards for Architecture in 2015

范曾艺术馆

项目地点 / 江苏省南通市
规模 / 7028 m²
业主 / 南通大学
建成时间 / 2014
主要用途 / 展示
获奖情况 / 2015 全国优秀工程勘察设计行业奖公建一等奖

Fan Zeng Art Museum

Location / Nantong, Jiangsu Province
Floor area / 7028 m²
Client / Nantong University
Completion / 2014
Function / Exhibition
Award / First Prize in Building Engineering from the Ministry of Education's Excellent Engineering Survey Award in 2015

山东美术馆

项目地点 / 山东省济南市
规模 / 52138 m²
业主 / 山东省文化厅
建成时间 / 2013
主要用途 / 美术馆
获奖情况 / 2015 年全国优秀工程勘察设计行业奖一等奖

Shandong Art Gallery

Location / Jinan, Shandong Province
Floor area / 52,138 m²
Client / Agency for Cultural Affairs, Shandong Province
Completion / 2013
Function / Gallery
Award / First Prize from the National Award of Excellent Engineering, Survey, and Design Industry in 2015

刘海粟美术馆

项目地点 / 上海市长宁区
规模 / 地上 10322 m²; 地下 2218 m²
业主 / 刘海粟美术馆
建成时间 / 2015
主要用途 / 美术馆展览
获奖情况 / 上海市建筑学会第五届建筑创作奖优秀奖

Liu Haisu Art Gallery

Location / Changning District, Shanghai
Floor area / Ground 10,322 m²; Underground 2218 m²
Client / Liu Haisu Art Museum
Completion / 2015
Function / Gallery, exhibition
Award / Excellence Award from the Architectural Society of Shanghai, Architectural Creation Awards

2015 米兰世博会·中国企业联合馆

项目地点 / 意大利米兰
规模 / 2000 m²
业主 / 上海米博投资发展有限公司
建成时间 / 2015
主要用途 / 展示
获奖情况 / 上海市建筑学会第六届建筑创作奖优秀奖

China Corporate United Pavilion, Expo 2015 Milano Italy

Location / Milan, Italy
Floor area / 2000 m²
Client / Shanghai Mibo Investment and Development Co., Ltd.
Completion / 2015
Function / Exhibition
Award / Excellence Award from the Architectural Society of Shanghai's Sixth Architectural Creation Awards

扬州广陵公共文化中心

项目地点 / 扬州市广陵新城
规模 / 97500 m²
业主 / 扬州美科置业有限公司
建成时间 / 建设中
获奖情况 / 2015 年上海市建筑学会第六届建筑创作奖优秀奖

Yangzhou Guangling Public Cultural Center

Location / Guangling New Town, Yangzhou
Floor area / 97,500 m²
Client / Yangzhou Meike Properties Co., Ltd.
Completion / On-going
Award / Excellence Award from the Architectural Society of Shanghai's Sixth Architectural Creation Awards in 2015

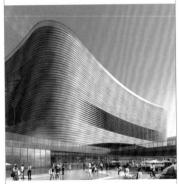

中国商业与贸易博物馆、义乌市美术馆

项目地点 / 浙江省义乌市
规模 / 地上 35000 m²;地下 5000 m²
业主 / 义乌市政府
建成时间 / 建设中
主要用途 / 博物馆、艺术馆
获奖情况 / 2015 上海市建筑学会第六届建筑创作奖佳作奖

China Business and Trade Museum, Yiwu Art Museum

Location / Yiwu, Zhejiang Province
Floor area / Ground 35,000 m²; Underground 5000 m²
Client / Yiwu Municipal Government
Completion / On-going
Function / Museum, art gallery
Award / Contribution of Excellence from the Architectural Society of Shanghai's Sixth Architectural Creation Awards in 2015

咸阳市市民文化中心

项目地点 / 陕西省咸阳市
规模 / 155000 m²
业主 / 咸阳市统建项目管理办公室
建成时间 / 建设中
主要用途 / 展示、观演、活动、档案
获奖情况 / 2013 年第五届上海市建筑学会建筑创作奖佳作奖

Xianyang Citizen Culture Center

Location / Xianyang, Shanxi Province
Floor area / 155,000 m²
Client / Xianyang Unified Construction Project Administration Office
Completion / On-going
Function / Display, performing, events, archives
Award / Contribution of Excellence from the Architectural Society of Shanghai's Fifth Architectural Creation Awards in 2013

遵义市娄山关红军战斗遗址陈列馆

项目地点 / 贵州省遵义市
规模 / 地上 685 m²;地下 5371 m²
业主 / 遵义市娄山关管理处
建成时间 / 2016
主要用途 / 博物馆展览
获奖情况 / 2015 年同济大学建筑设计研究院(集团)有限公司建筑创作奖一等奖

Zunyi Loushan Pass Red Army Battle Site Museum

Location / Zunyi, Guizhou Province
Floor area / Ground 685 m²; Underground 5371 m²
Client / Loushanguan Administration Office of Zunyi City
Completion / 2016
Function / Convention and exhibition
Award / First Prize from TJAD's Architecture Creation Awards in 2015

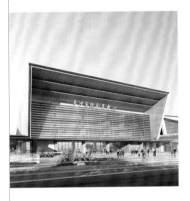

长沙国际会展中心

项目地点 / 湖南省长沙市
规模 / 445100 m²
业主 / 湖南长沙会展中心投资有限责任公司
建成时间 / 2016
主要用途 / 会展

Changsha International Convention and Exhibition Center

Location / Changsha, Hunan Province
Floor area / 445,100 m²
Client / Changsha International Convention Center Investment Co., Ltd., Hunan Province
Completion / 2016
Function / Convention and exhibition

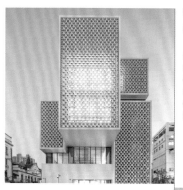

Wuzhong District Dongwu Culture Center

Location / Suzhou, Jiangsu Province
Floor area / 143,519 m²
Client / Suzhou Guorong Cultural Development Co., Ltd.
Completion / 2016
Function / Office, competition

上海棋院

项目地点 / 上海市静安区
规模 / 12424 m²
业主 / 上海棋院
建成时间 / 建设中
主要用途 / 办公、比赛

Shanghai Chess Institute

Location / Jiang'an District, Shanghai
Floor area / 12,424 m²
Client / Shanghai Qiyuan
Completion / On-going
Function / Office, competition

福州城市发展展示馆

项目地点 / 福建省福州市
规模 / 53300 m²
业主 / 福建省二建建设集团有限公司
建成时间 / 2014
主要用途 / 文化中心
获奖情况 / 2013 年第五届上海市建筑学会建筑创作奖佳作奖

Fuzhou Urban Development Exhibition Hall

Location / Fuzhou, Fujian Province
Floor area / 53,300 m²
Client / The Second Construction (Group) Co., Ltd., Fujian Province
Completion / 2014
Function / Display, office, management and service center
Award / Contribution of Excellence from the Architectural Society of Shanghai's Fifth Architectural Creation Awards in 2013

吴中区东吴文化中心

项目地点 / 江苏省苏州市
规模 / 143519 m²
业主 / 苏州国融文化发展有限公司
建成时间 / 2016
主要用途 / 文化中心

上海交响乐团迁建工程

项目地点 / 上海市徐汇区
合作设计单位 / 矶崎新设计工作室
规模 / 地上 5274 m²；地下 14676 m²
业主 / 上海交响乐团
建成时间 / 2013
主要用途 / 音乐演出
获奖情况 / 2015 年度上海市优秀工程勘察设计一等奖

Relocation of the Shanghai Symphony Orchestra

Location / Xuhui District, Shanghai
Collaborators / Arata Isozaki & Associates
Floor area / Ground 5274 m²; Underground 14,676 m²
Client / Shanghai Symphony Orchestra
Completion / 2013
Function / Music performing
Award / First Prize from the Award of Excellent Engineering, Survey and Design Industry of Shanghai in 2015

嘉定新城保利大剧院

项目地点 / 上海市嘉定新城
合作设计单位 / 安藤忠雄建筑研究所
规模 / 地上 35840 m²；地下 20064 m²
业主 / 上海保利茂佳房地产开发有限公司
建成时间 / 2014
主要用途 / 大型剧院
获奖情况 / 2015 年上海市优秀工程勘察设计一等奖

Jiading New Town Poly Grand Theater

Location / Jiading New Town, Shanghai
Collaborators / Tadao Ando Architect & Associates
Floor area / Ground 35,840 m²; Underground 20,064 m²
Client / Shanghai Poly Maojia Real Estate Development Co., Ltd.
Completion / 2014
Function / Theater
Award / First Prize from the Award of Excellent Engineering, Survey and Design Industry of Shanghai in 2015

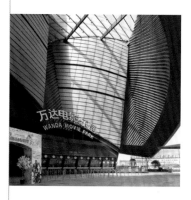

武汉电影乐园

项目地点 / 湖北省武汉市
合作设计单位 / Stufish Entertainment Architects, Forrec Ltd
规模 / 地上 48934 m²；地下 52617 m²
业主 / 武汉万达东湖置业有限公司
建成时间 / 2013
主要用途 / 电影体验、娱乐、商业、餐饮
获奖情况 / 2015 年度上海市优秀工程勘察设计三等奖

Wanda Wuhan Movie Park

Location / Wuhan, Hubei Province
Collaborators / Stufish Entertainment Architects, Forrec Ltd
Floor area / Ground 48,934 m²; Underground 52,617 m²
Client / Wuhan Wanda Donghu Properties Co., Ltd.
Completion / 2013
Function / Movie, recreation, commercial, catering
Award / Third Prize from the Award of Excellent Engineering, Survey and Design Industry of Shanghai in 2013

上海市虹口区海南路 10 号地块项目

项目地点 / 上海市虹口区
合作设计单位 / 隈研吾建筑都市设计事务所
规模 / 地上 68522.51 m²；地下 26554 m²
业主 / SOHO 中国有限公司
建成时间 / 2015
主要用途 / 办公及其配套商业
获奖情况 / 2016 年中国最佳高层建筑奖

Plot No. 10, Hainan Road, Hongkou District, Shanghai

Location / Hongkou District, Shanghai
Collaborators / Kengo Kuma & Associates
Floor area / Ground 68,522.51 m²;
Underground 26,554 m²
Client / SOHO China Co., Ltd.
Completion / 2015
Function / Office and supporting commercial facilities
Award / The Best Chinese High-Rise Building of 2016

上海市城市建设投资开发总公司企业自用办公楼

项目地点 / 上海市杨浦区
规模 / 21968 m²
业主 / 上海市城市建设投资开发总公司
建成时间 / 2013
主要用途 / 办公
获奖情况 / 2015 年上海市优秀工程勘察设计二等奖

Office Building of Shanghai City Construction and Investment Corporation

Location / Yangpu District, Shanghai
Floor area / 21,968 m²
Client / Shanghai Urban Construction & Investment Corporation
Completion / 2013
Function / Office
Award / Second Prize from the Award of Excellent Engineering, Survey and Design Industry of Shanghai in 2015

平凉街道 22 街坊项目

项目地点 / 上海市杨浦区
合作设计单位 / GMP
规模 / 208361 m²
业主 / 上海盛冠房地产有限公司
建成时间 / 建设中
主要用途 / 办公、商业
获奖情况 / 中华人民共和国住房和城乡建设部"二星级绿色建筑设计标识"

Project of No. 22 Pingliang Street

Location / Yangpu District, Shanghai
Collaborators / GMP
Floor area / 208,361 m²
Client / Shanghai Shengguan Real Estate Co., Ltd.
Completion / On-going
Function / Office, commercial use
Award / Authentication of Two-Star Green Building Design from Chinese MOHURD

同济大厦 A 楼

项目地点 / 上海市杨浦区
规模 / 36982 m²
业主 / 同济大学
建成时间 / 2011
主要用途 / 办公、教学
获奖情况 / 2013 全国优秀工程勘察设计行业奖二等奖

Building A of Tongji Tower

Location / Yangpu District, Shanghai
Floor area / 36,982 m²
Client / Tongji University
Completion / 2011
Function / Office, teaching
Award / Second Prize of the National Award of Excellent Engineering, Survey and Design Industry in 2013

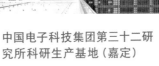

中国电子科技集团第三十二研究所科研生产基地（嘉定）

项目地点 / 上海市嘉定区
规模 / 地上 12128 m²；地下 32958 m²
业主 / 中国电子科技集团第三十二研究所
建成时间 / 2016
主要用途 / 办公、科研、员工餐厅与活动、生产、检测

Production Base of the 32nd Research Institute of the China Electronics Technology Group (Jiading)

Location / Jiading District, Shanghai
Floor area / Ground 12,128 m²; Underground 32,958 m²
Client / No.32 Research Institute of CETC
Completion / 2016
Function / Office, scientific research, staff dining and activities, production, detection

英特宜家无锡购物中心

项目地点 / 江苏省无锡市
规模 / 地上 223228 m²；地下 2539 m²
业主 / 英特宜家购物中心（中国）管理有限公司
建成时间 / 2014
主要用途 / 商业

Wuxi Inter IKEA Shopping Center

Location / Wuxi, Jiangsu Province
Floor area / Ground 223,228 m²; Underground 2539 m²
Client / Wuxi Inter IKEA Center (China) Management Co., Ltd.
Completion / 2014
Function / Commercial use

建成时间 / 2015
主要用途 / 商业、办公

L-FERG Zhabei Project, Shanghai

Location / Zhabei District, Shanghai
Collaborators / UN Studio
Floor area / 348,337 m²
Client / Liyida Commercial Properties (Shanghai) Co., Ltd.
Completion / 2015
Function / Commercial use, office

Zhengzhou 27 New Tower

Location / Zhengzhou, Henan Province
Floor area / 451,710 m²
Client / Greenland Group
Completion / On-going
Function / Commercial use, office, 5-star hotel
Award / First Prize from TJAD's Architecture Creation Awards in 2014

黄河口生态旅游区游船码头

项目地点 / 山东省东营市
规模 / 5100 m²
业主 / 东营市旅游开发有限公司
建成时间 / 2010
主要用途 / 候船、观光、文化、展示
获奖情况 / 2011 中国建筑学会第六届建筑创作奖佳作奖

Cruise Terminal in the Ecological Tourism Zone of the Yellow River Estuary

Location / Dongying, Shandong Province
Floor area / 5100 m²
Client / Dongying Tourism Development Co., Ltd.
Completion / 2010
Function / Waiting, tourism, culture, display
Award / Contribution of Excellence from the Architectural Society of China's Sixth Architectural Creation Awards in 2011

英特宜家武汉购物中心

项目地点 / 湖北省武汉市
合作设计单位 / Building Design Partnership
规模 / 248128.67 m²
业主 / 武汉英特宜家置业有限公司
建成时间 / 2015
主要用途 / 商业

Wuhan Inter IKEA Shopping Center

Location / Wuhan, Hubei Province
Collaborators / Building Design Partnership
Floor area / 248,128.67 m²
Client / Wuhan Inter IKEA Properties Co., Ltd.
Completion / 2015
Function / Commercial use

温州机场交通枢纽综合体

项目地点 / 浙江省温州市
规模 / 300416 m²
业主 / 温州机场集团有限公司
建成时间 / 建设中
主要用途 / 商业服务、多功能城市综合体

Wenzhou Airport Traffic Hub Complex

Location / Wenzhou, Zhejiang Province
Floor area / 300,416 m²
Client / Wenzhou Airport Group Co., Ltd.
Completion / On-going
Function / Commercial service, multi-functional complex

郑州 27 新塔项目

项目地点 / 河南省郑州市
规模 / 451710 m²
业主 / 绿地集团
建成时间 / 设计中
主要用途 / 商业、办公、五星级酒店
获奖情况 / 2014 年同济大学建筑设计研究院（集团）有限公司建筑创作一等奖

昆明滇池国际会展中心 4 号地块（主塔）

项目地点 / 云南省昆明市
规模 / 155812.7 m²
业主 / 云南新世纪滇池国际文化旅游会展投资有限公司
建成时间 / 设计中
主要用途 / 酒店、办公
获奖情况 / 第十一届同济大学建筑设计研究院（集团）有限公司建筑创作二等奖

Plot No. 4 of Kunming Dianchi International Convention and Exhibition Center (main tower)

Location / Kunming, Yunnan Province
Floor area / 155,812.7 m²
Client / New Century Dianchi International Cultural Tourism and Exhibition Investment Co., Ltd., Yunnan Province
Completion / On-going
Function / Hotel, office
Award / Second Prize from TJAD's 11th Architecture Creation Awards

利福上海闸北项目

项目地点 / 上海市闸北区
合作设计单位 / 日本设计、UN Studio
规模 / 348337 m²
业主 / 利怡达商业置业（上海）有限公司

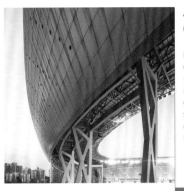

Jining Olympics Sports Center

Location / Jining, Shandong Province
Collaborators / SCAU Architects (Shanghai)
Floor area / 175,602 m²
Client / preparatory Team of the 23rd Shandong Provincial Sports Games
Completion / 2014
Function / Sports architecture

遂宁市体育中心

项目地点 / 四川省遂宁市
规模 / 79741 m²
业主 / 遂宁市河东开发建设投资有限公司
建成时间 / 2014
主要用途 / 酒店、办公
获奖情况 / 2015年度上海市优秀工程勘察设计一等奖

Suining Sports Center

Location / Suining, Sichuan Province
Floor area / 79,741 m²
Client / Suining Hedong Development and Construction Investment Co., Ltd.
Completion / 2014
Function / Sports architecture
Award / First Prize from the Award of Excellent Engineering, Survey and Design Industry of Shanghai in 2015

济宁奥体中心

项目地点 / 山东省济宁市
合作设计单位 / 斯构莫尼建筑设计咨询（上海）有限公司
规模 / 175602 m²
业主 / 济宁市第二十三届省运会筹备工作领导小组
建成时间 / 2014
主要用途 / 体育建筑

常熟市体育中心体育馆

项目地点 / 江苏省常熟市
规模 / 32249 m²
业主 / 常熟市城市经营投资有限公司
建成时间 / 2011
主要用途 / 体育建筑
获奖情况 / 2015年度全国优秀工程勘察设计行业奖二等奖

Gymnasium of Changshu Sports Center

Location / Changshu, Jiangsu Province
Floor area / 32,249 m²
Client / Changshu Urban Operation and Investment Co., Ltd.
Completion / 2011
Award / Second Prize from the National Awards for Excellence in the Engineering, Survey, and Design Industry in 2015

上海崇明体育训练基地

项目地点 / 上海市崇明区
规模 / 189708 m²
业主 / 上海体育职业学院
建成时间 / 2018
主要用途 / 体育训练、教学、医疗

Shanghai Chongming Sports Training Base

Location / Chongming District, Shanghai
Floor area / 189,708 m²
Client / Shanghai Vocational College of Sports
Completion / 2018
Function / Training, teaching, medical treatment

改建铁路宁波站改造工程

项目地点 / 浙江省宁波市
合作设计单位 / 铁道第三勘察设计院集团有限公司
规模 / 119600 m²
业主 / 上海铁路局宁波铁路枢纽工程建设指挥部
建成时间 / 2013
主要用途 / 大型综合交通枢纽
获奖情况 / 2015年全国优秀工程勘察设计行业奖三等奖

Reconstruction of Ningbo Railway Station

Location / Ningbo, Zhejiang Province
Collaborators / Ministry of Railways 3rd Survey and Design Institute Group
Floor area / 119,600 m²
Client / Ningbo Railway Pivotal Project Construction Headquarters of Shanghai Railway Bureau
Completion / 2013
Function / Comprehensive transportation hub
Award / Third Prize from the National Awards for Excellence in the Engineering, Survey, and Design Industry in 2015

哈大客专大连北站站房工程

项目地点 / 辽宁省大连市
合作设计单位 / 铁道第三勘察设计院集团有限公司
规模 / 160495 m²
业主 / 沈阳铁路局新大连站工程建设指挥部
建成时间 / 2012
主要用途 / 大型综合交通枢纽
获奖情况 / 2015年度全国优秀工程勘察设计行业奖二等奖

Station Building of Dalian North Station of the Harbin-Dalian Passenger Line

Location / Daliang, Liaoning Province
Collaborators / Ministry of Railways 3rd Survey and Design Institute Group
Floor area / 160,495 m²
Client / New Dalian Railway Station Project Construction Headquarters of Shenyang Railway Bureau
Completion / 2012
Function / Public transportation architecture
Award / Second Prize from the National Awards for Excellence in the Engineering, Survey, and Design Industry in 2015

建成时间 / 2016
主要用途 / 交通枢纽综合体

Haikou Coach Terminal

Location / Haikou, Hainan Province
Collaborators / Hainan South Architecture Design Ltd.
Floor area / 64,580 m²
Client / Hainan Haikou Bus Transportation Group Plc.
Completion / 2016
Function / Transportation hub complex

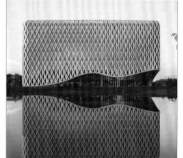

主要用途 / 图书馆
获奖情况 / 2015 年度上海市优秀工程勘察设计一等奖

Library of Zhejiang College of Tongji University

Location / Jiaxing, Zhejiang Province
Floor area / Ground 26875 m²; Underground 3131 m²
Client / Zhejiang College of Tongji University
Completion / 2014
Function / Library
Award / First Prize from the Award of Excellent Engineering, Survey and Design Industry of Shanghai in 2015

兰州西站站房工程

项目地点 / 甘肃省兰州市
规模 / 260000 m²
业主 / 兰州铁路局兰州枢纽工程建设指挥部
建成时间 / 2014
主要用途 / 铁路客运站房
获奖情况 / 第九届中国建筑学会优秀建筑结构设计二等奖

Station Building of Lanzhou West Railway Station

Location / Lanzhou, Gansu Province
Floor area / 260,000 m²
Client / Lanzhou Pivotal Project Construction Headquarters of Lanzhou Railway Bureau
Completion / 2014
Function / Railway station building
Award / Second Prize from the Architectural Society of China's Ninth Building Structure Design Awards

上海吴淞口国际邮轮码头后续工程

项目地点 / 上海市宝山区
合作设计单位 / 中交第三航务工程勘察设计院有限公司
规模 / 55408 m²
业主 / 上海吴淞口国际邮轮港发展有限公司
建成时间 / 建设中
主要用途 / 国际邮轮客运中心

北京建筑大学新校区图书馆

项目地点 / 北京市大兴区
规模 / 地上 25212 m²; 地下 10599 m²
业主 / 北京建筑大学
建成时间 / 2014
主要用途 / 图书馆
获奖情况 / 上海市建筑学会第六届建筑创作佳作奖

Library in the New Campus of Beijing University of Civil Engineering and Architecture

Location / Daxing District, Beijing
Floor area / Ground 25,212 m²; Underground 10,599 m²
Client / Beijing University of Civil Engineering and Architecture
Completion / 2014
Function / Library
Award / Contribution of Excellence from the Architectural Society of Shanghai's Sixth Architectural Creation Awards

西北工业大学长安校区图书馆

项目地点 / 陕西省西安市
规模 / 53.631 m²
业主 / 西北工业大学
建成时间 / 2013
主要用途 / 图书馆
获奖情况 / 2015 年教育部优秀建筑设计一等奖

Follow-up Project of the Shanghai Wusongkou International Cruise Terminal

Location / Baoshan District, Shanghai
Collaborators / Chinese Transportation 3rd Harbor Engineering Survey and Design Institute Ltd.
Floor area / 55,408 m²
Client / Shanghai Wusongkou International Cruise Port Development Co., Ltd.
Completion / On-going
Function / International cruise transportation center

Library of the Chang'an Campus of Northwestern Polytechnical University

Location / Xi'an, Shanxi Province
Floor area / 53631 m²
Client / Northwestern Polytechnical University
Completion / 2013
Function / Library
Award / First Prize from the Ministry of Education's Awards for Excellence in Architectural Design in 2015

海口汽车客运总站

项目地点 / 海南省海口市
合作设计单位 / 海南南方建筑设计有限公司
规模 / 64580 m²
业主 / 海南海汽运输集团股份有限公司

同济大学浙江学院图书馆

项目地点 / 浙江省嘉兴市
规模 / 地上 26875 m²; 地下 3131 m²
业主 / 同济大学浙江学院
建成时间 / 2014

南开大学津南校区

项目地点 / 天津市海河教育园区
规模 / 核心教学区 112030 m²；文科学院组团 76022 m²；新兴学科组团 51858 m²；体育馆 23189 m²；大学生活动中心 10947 m²
业主 / 南开大学
建成时间 / 2015
主要用途 / 教育
获奖情况 / 2013 年第五届上海市建筑学会建筑创作奖佳作奖

Jinnan Campus of Nankai University

Location / Haihe Education Park, Tianjin
Floor area / Core Teaching Area 112,030 m²;
Cluster of Colleges of Liberal Arts 76,022 m²;
Cluster of New Subjects; Stadium 23,189 m²;
Students Activities Center 10,947 m²
Client / Nankai University
Completion / 2015
Function / Education
Award / Contribution of Excellence from the Architectural Society of Shanghai's Fifth Architectural Creation Awards in 2013

浦江镇江柳路幼儿园

项目地点 / 上海市闵行区
规模 / 15329 m²
业主 / 上海浦江镇投资发展有限公司
建成时间 / 2015
主要用途 / 二十班日托幼儿园
获奖情况 / 2013 年第九届同济大学建筑设计研究院（集团）有限公司建筑创作奖三等奖

Pujiang Town Jiangliu Road Kindergarten

Location / Minhang District, Shanghai
Floor area / 15,329 m²
Client / Shanghai Pujiang Town Investment and Development Co., Ltd.
Completion / 2015
Function / 20 class daycare kindergarten
Award / Third Prize from TJAD's Ninth Architecture Creation Awards in 2013

上海市委党校二期工程（教学楼、学员楼）

项目地点 / 上海市徐汇区
规模 / 地上 28447 m²；地下 8426 m²
业主 / 中共上海市委党校
建成时间 / 2011
主要用途 / 教学、宿舍
获奖情况 / 2015 年度全国优秀工程勘察设计行业奖绿色建筑一等奖

The School of the Shanghai Municipal Communist Party of China, Phase II (Teaching Building, Student Building)

Location / Xuhui District, Shanghai
Floor area / Ground 28,447 m²;
Underground 8426 m²
Client / Shanghai Municipal Party School
Completion / 2011
Function / Teaching, dormitory
Award / First Prize from the National Award of Excellent Engineering, Survey, and Design Industry in Green Buildings in 2015

新江湾城中福会幼儿园

项目地点 / 上海市杨浦区
规模 / 8179 m²
业主 / 上海市城市建设投资开发总公司
建成时间 / 2009
主要用途 / 17 班幼儿园
获奖情况 / 2013 中国勘察设计协会行业奖公建一等奖

New Jiangwan Town Zhongfuhui Kindergarten

Location / Yangpu District, Shanghai
Floor area / 8179 m²
Client / Shanghai Urban Construction & Investment Corporation
Completion / 2009
Function / 17 class kindergarten
Award / First Prize in Public Buildings from the China Exploration & Design Association's Industry Awards in 2013

苏州大学附属第一医院平江分院

项目地点 / 江苏省苏州市
合作设计单位 / 株式会社山下设计
规模 / 地上 132182 m²；地下 53,792 m²
业主 / 苏州大学附属第一医院
建成时间 / 2015
主要用途 / 医疗建筑

The First Affiliated Hospital of Suzhou University, Pingjiang Branch

Location / Suzhou, Jiangsu Province
Collaborators / Yamashita Sekkei Inc
Floor area / Ground 132,182 m²;
Underground 53,792 m²
Client / The Fist Affiliated Hospital of Suzhou University
Completion / 2015
Function / Medical architecture

苏州市第九人民医院

项目地点 / 江苏省苏州市
规模 / 地上 224283 m²；地下 83,369 m²
业主 / 苏州市第九人民医院
建成时间 / 建设中
主要用途 / 医疗建筑
获奖情况 / 上海市建筑学会第六届建筑创作奖佳作奖

The Ninth People's Hospital of Suzhou

Location / Suzhou, Jiangsu Province
Floor area / Ground 224,283 m²;
Underground 83,369 m²
Client / The Ninth People's Hospital of Suzhou
Completion / On-going
Function / Medical architecture
Award / Contribution of Excellence from the Architectural Society of Shanghai's Sixth Architectural Creation Awards

Xuhui District Southern Medical Center

Location / Xuhui District, Shanghai
Collaborators / Perkins+Will
Floor area / 212,800 m²
Client / Xuhui District Center Hospital of Shanghai; Shanghai Xuhui District Land Reservation Co., Ltd.
Completion / On-going
Function / General hospital, specialized hospital, scientific research, office

Jimei Marriott Hotel

Location / Xiamen, Fujian Province
Floor area / 99,815 m²
Client / Xiamen Xinglin Construction & Development Co., Ltd.
Completion / On-going
Function / Hotel

上海市第一人民医院改扩建工程

项目地点 / 上海市虹口区
规模 / 48130 m²
业主 / 上海市第一人民医院
建成时间 / 2016
主要用途 / 医疗建筑

Expansion and Reconstruction of the First People's Hospital of Shanghai

Location / Hongkou District, Shanghai
Floor area / 48,130 m²
Client / The First People's Hospital of Shanghai
Completion / 2016
Function / Medical architecture

青岛岭海温泉大酒店

项目地点 / 山东省青岛市
规模 / 129600 m²
业主 / 青岛长基置业有限公司
建成时间 / 2015
主要用途 / 酒店

Qingdao Linghai Springs Hotel

Location / Qingdao, Shandong Province
Floor area / 129,600 m²
Client / Qingdao Changji Properties Co., Ltd.
Completion / 2015
Function / Hotel

国家机关事务管理局第二招待所翻扩建工程

项目地点 / 北京市西城区
规模 / 99980 m²
业主 / 国务院第二招待所
建成时间 / 2016
主要用途 / 会议、住宿

Expansion of the Second Guest House of the National Affairs Management Bureau

Location / Xicheng District, Beijing
Floor area / 99,980 m²
Client / The Second Guest House of State Council
Completion / 2016
Function / Convention, accommodation

昆明花之城豪生国际大酒店

项目地点 / 云南省昆明市
合作设计单位 / 云南省设计院
规模 / 250664 m2
业主 / 昆明怡美天香置业有限公司
建成时间 / 2015
主要用途 / 酒店、商业、植物园

Howard Johnson Plaza Hotel, Flower City, Kunming

Location / Kunming, Yunnan Province
Collaborators / Design Institute of Yunnan Province
Floor area / 250,664 m²
Client / Kunming Yimei Tianxiang Properties Co., Ltd.
Completion / 2015
Function / Hotel, commercial use, botanic garden

徐汇区南部医疗中心

项目地点 / 上海市徐汇区
合作设计单位 / Perkins+Will
规模 / 212800 m²
业主 / 上海市徐汇区中心医院；上海徐汇土地储备有限公司
建成时间 / 建设中
主要用途 / 综合医院、专科医院、科研、办公

集美万豪酒店

项目地点 / 福建省厦门市
规模 / 99815 m²
业主 / 厦门市杏林建设开发有限公司
建成时间 / 建设中
主要用途 / 酒店

黄山元一柏庄国际旅游体验中心－国际酒店

项目地点／安徽省黄山市
合作设计单位／MAP architecture & planning
规模／80982 m²
业主／安徽黄山元一柏庄投资发展有限公司
建成时间／2012
主要用途／五星级酒店及贵宾别墅

International Hotel of Huangshanyuanyi Bozhuang International Tourism Experience Center

Location / Huangshan, Anhui Province
Collaborators / MAP architecture & planning
Floor area / 80,982 m²
Client / Huangshan Yuanyi Baizhuang Investment and Development Co., Ltd., Anhui Province
Completion / 2012
Function / 5-star hotel, VIP villas

上海鞋钉厂改建项目（原作设计工作室）

项目地点／上海市杨浦区
规模／1000 m²
业主／同济大学建筑设计研究院（集团）有限公司
建成时间／2013
主要用途／办公
获奖情况／2015亚洲建筑师协会建筑奖荣誉提名奖

Reconstruction of the Shanghai Hobnail Factory (Original Design Studio)

Location / Yangpu District, Shanghai
Floor area / 1000 m²
Client / TJAD
Completion / 2013
Function / Office
Award / Nominated for ARCASIA's Architecture Award in 2015

同济大学设计创意学院

项目地点／上海市杨浦区
规模／9007 m²
业主／同济大学
建成时间／2013
主要用途／教学、办公
获奖情况／2015年上海市建筑学会第六届建筑创作奖优秀奖

College of Design and Innovation, Tongji University

Location / Yangpu District, Shanghai
Floor area / 9007 m²
Client / Tongji University
Completion / 2013
Function / Teaching, office
Award / Excellence Award from the Architectural Society of Shanghai's Sixth Architectural Creation Awards in 2015

延安中路816号（解放日报社）

项目地点／上海市静安区
规模／5370 m²
业主／上海文新经济发展有限公司
建成时间／2016
主要用途／办公

No. 816 Yan'an Middle Road (Jiefang Daily)

Location / Jing'an District, Shanghai
Floor area / 5370 m2
Client / Shanghai Wenxin Economic Development Co., Ltd.
Completion / 2016
Function / Office

同济大学博物馆

项目地点／上海市杨浦区
规模／4469 m²
业主／同济大学
建成时间／2013
主要用途／博物馆展览
获奖情况／2014年度中国建筑学会建筑创作奖——建筑保护与再利用类银奖

Museum of Tongji University

Location / Yangpu District, Shanghai
Floor area / 4469 m²
Client / Tongji University
Completion / 2013
Function / Exhibition
Award / Silver Medal in Architectural Conservation and Reuse from the Architectural Society of China's Architectural Creation Awards in 2014

同济大学建筑与城规学院D楼改建项目

项目地点／上海市杨浦区
规模／6440 m²
业主／同济大学
建成时间／2010
主要用途／博物馆展览
获奖情况／2014年度中国建筑学会建筑创作奖——建筑保护与再利用类银奖

Reconstruction of Building D of the College of Architecture and Urban Planning, Tongji University

Location / Yangpu District, Shanghai
Floor area / 6440 m²
Client / Tongji University
Completion / 2010
Function / Classroom, laboratory, auditorium
Award / Silver Medal in Architectural Conservation and Reuse from the Architectural Society of China's Architectural Creation Awards in 2014

本书中的信息和图片均由同济大学建筑设计研究院（集团）有限公司准备并提供。感谢所有参与本书信息采集、图片拍摄、材料整理的工作人员。虽然在出版过程中我们已尽量确保内容的准确性，但难免有疏漏之处。欢迎读者来信更正我们的错误与遗漏，联系信箱：5yyj@tjadri.com。

图书在版编目(CIP)数据

TJAD2012－2017作品选：汉、英／丁洁民主编．—桂林：广西师范大学出版社，2017.6
ISBN 978－7－5495－9601－0

Ⅰ．①T… Ⅱ．①丁… Ⅲ．①建筑设计－作品集－中国－现代 Ⅳ．①TU206

中国版本图书馆CIP数据核字(2017)第065486号

出 品 人：刘广汉
责任编辑：肖　莉　季　慧
版式设计：张　晴
广西师范大学出版社出版发行
（广西桂林市中华路22号　　邮政编码：541001）
（网址：http://www.bbtpress.com）
出版人：张艺兵
全国新华书店经销
销售热线：021－31260822－882/883
上海利丰雅高印刷有限公司印刷
（上海庆达路106号　邮政编码：201200）
开本：889mm×1 270mm　　1/16
印张：22　　　　　　　字数：20千字
2017年6月第1版　　2017年6月第1次印刷
定价：328.00元

如发现印装质量问题，影响阅读，请与印刷单位联系调换。